INSECT PEST PREDATORS

T.V. Sathe
Y.A. Bhosale

2020
Daya Publishing House®
A Division of
Astral International Pvt. Ltd.
New Delhi - 110 002

First Published, 2001
Reprinted, 2020

ISBN: 978-81-7035-235-8

Publisher s note:

Every possible effort has been made to ensure that the information contained in this book is accurate at the time of going to press, and the publisher and author cannot accept responsibility for any errors or omissions, however caused. No responsibility for loss or damage occasioned to any person acting, or refraining from action, as a result of the material in this publication can be accepted by the editor, the publisher or the author. The Publisher is not associated with any product or vendor mentioned in the book. The contents of this work are intended to further general scientific research, understanding and discussion only. Readers should consult with a specialist where appropriate.

Every effort has been made to trace the owners of copyright material used in this book, if any. The author and the publisher will be grateful for any omission brought to their notice for acknowledgement in the future editions of the book.

Published by : **Daya Publishing House®**
A Division of
Astral International Pvt. Ltd.
– ISO 9001:2015 Certified Company –
4736/23, Ansari Road, Darya Ganj
New Delhi-110 002
Ph. 011-43549197, 23278134
E-mail: info@astralint.com
Website: www.astralint.com

Digitally Printed at : **Replika Press Pvt. Ltd.**

Preface

Recently, increasing efforts have been directed towards Biological pest control since it is very effective alternative for chemical control. Pesticides, however, act effectively against pests but lead to many serious problems like air and water pollution, health hazards, killing of beneficial insects, pest resistance, secondary pest outbreak, pest resurgence, stimulation in the rate of reproduction in certain insects, disturbance in ecocycle, etc.

In the biocontrol programme of insect pests, predators are widely used. The coccinelids (lady bird beetles) are predaceus insects and very effective biocontrol agents of various insect pests. Lace wings, pentatomids, mirids, syrphids are also important predators of insect pests.

In the text, 19 coccinclid species have been described, the methods of collection, preservation and rearing also been described. In addition, biology of a coccinelid, predator pray relationships, Pest predator index and biocontrol programmes designed with predators in India and abroad have also been incorporated in the text. The text explores several avenues for future research in the field of biological pest control by predators. Authors are thankful to Shivaji University, Kolhapur (M.S.) for providing facilities to this work. We feel and hope that the book would be useful and stimulatory for farmers, students, teachers and researchers in basic and applied field pest management both from India and abroad.

T.V. Sathe

Y.A. Bhosale

Contents

1

INTRODUCTION

Indian agriculture is predominantly characterised by the cultivation of a wide variety of food and non-food crops. The food crops refer to cereals, pulses, oilseeds, vegetables, fruits and others (Singh and Sadhu, 1986). In last forty years, croping intensity increåsed from 1.11 to 1.36 (Paroda, 1993). It has been estimated that out of the total precipitation of around 400 million hectare meters in the country, the suface water availability is about 178 million hectare meters. There are several significant achievements in different crops relating to breeding crops either genetically higher in productivity or possessing resistance to various biotic and abiotic traces. The compound growth rate of various crops/ commodities during the last forty years has been fairly impressive but the expected yield have not achieved so far because of the damage caused by the insects to the crop.

A very large number of pests inflict the damage to the stems, leaves, flowers, fruits, seeds and roots of agricultural and forest plants. Many soil inhibiting insects damage roots and many kill the plants causing serious losses. As per the estimate made in 1970, annual crop losses exceeding Rs. 6,000 crores are inflicted by pests, of which insects cause 20 per cent (Paroda, 1993). The protection of crops from various kinds of pest remains a pressing problem and more increased in certain high yielding hybrid varieties.

The varieties of techniques and tactics are adopted in IPM (Integrated Pest Management) now a days. The management strategy for each type of pest being different, control operations are complex and highly technical. In IPM, a series of steps are followed to keep the population level below the economic threshold and demands a judicious use of pesticides, disease free seeds of resistant varieties, biological supression of pest population, usage of pheromones, timely spray of pesticides, etc. It is decision making process to determine if, when, where and what strategy and/or mix of tactics should use (Smith, 1980). However, at this time no other measures proved to be widely successful as chemical control in pest management programme.

Pesticide usage is comparatively new in India. From a mere 4000 metric tonnes of technical grade material in 1950, the usage has grown at a rate of 7.5% per annum and would exceed 82 metric tonnes during crop year 1991 (Parada, 1993). In the majority of cases, insecticides when spread on crops are reported to reduce the populations of phytophagous insects. However, several insecticides, used on orchard and field crops, bring about tremendous increase of the insects against which these have been applied or of other phytophagous species present only in small number at the time of treatment. From a state of over kill of pesticides, the approach has moved towards ecologically sound integrated pest management and biological control has a ecological base. The pesticides increase the yield of crops but the use of them in an indiscriminate way bring pest problems in a severe form and total devastation of crops have been noticed. The improper use of pesticides lead to the serious problems like air and water pollution, health hazards, physical and physiological changes in the soil, detrimental effects on beneficial insects (predators, parasitoids, honeybees, etc.), destruction of natural balance and ecological cycle, development of resistant variety of pests, pest resurgence, secondary pest outbreak, etc. Viewing from the above facts there is an urgent need to develop an alternative newer approach to manage the pest that flare-up when the ecosystems are disturbed due to the aggressive and unscientific methods adopted by farmers. In this contest, biological control is living weapon over chemical control, which is modern and widely accepted technique in the world.

Biological control is, "the action of parasitoids, predators and pathogens in maintaining other organisms density at a lower average than would occur in their absence" (DeBach, 1964). Coppel and Mertins 1977 defined biological control as, "the use or encouragement by man of living organisms or their productions for the population reduction of pest insects of biological control."

The predators are lions, tigers and hawks of the insect world and act as potent bio-control agents. The predators are defined as, "an organism which directly attacks, kills and eats one of the other species (prey)". In general, predators are categorised in two groups by taking into account of their feeding habit, predators stinging and predators biting and sucking. In the later type, a very large number of insects living in soil and under bark notably are beetles and smaller bugs. Generally predaotrs are larger than their prey, they consume more than one préy individual by moving around freely in both their immature and adult stages.

The predators are scattered in about 167 families of 14 orders of class Insecta. Coleoptera, Neuroptera, Hymenoptera, Diptera and Hemiptera contain a very large number of individuals which frequently determine the populations of pest species. Among the predaceous orders, Coleoptera ranks first. The family Coccinellidae of the order Coleoptera is exclusively predaceous except Epilachna beetles. Coccinellids, lady bugs, lady birds or preferably lady beetles have been respected through the centuries, as predators from this family which contains about 4000 species in the world.

The inability of insecticides to solve the permanent problem of the pests has increased the support and interest in biological control. World organizations like CIBC, CAB, CABI, IOBC, etc. are completely devoted to research in biological control. These organisations provide effective services to world agriculture to co-ordinate and administer agricultural informations, identification, biological control and other services. The commercial insectaries have made remarkable progress in these aspects and are largely responsible for transferring biological technology from lab to land in several countries (Manjunath, 1992). In U.S.A. commercial production of predators and parasitoids seems to be quite popular. There are atleast 50 commercial insectaries, which annually produce some 3000 million *Trichogramma* spp., 203 million *Aphytis melinus* DeBach, 26 million *Cryptolaemus montrouzieri* Mulsant and 18 million *Chrysoperla carnea* Stephens for control of various lepidopteran pests, scale insects, mealy bugs and other soft bodied sucking pests. In Russia, more than 10 biological factories are engaged in mass production of *Trichogramma* spp., and about 15 species of parasitoids and predators are produced and released in about 10 million hectares for control of variety of pests. In China, several species of *Trichogramma* are mass reared and used on a variety of crops, of which cotton alone comprised some 6,80,000 hectare annually. In Britain 170 million *Encarsia* are produced annually for control of white flies infesting vegetable crops in glass houses. DeBach (1964) reported more than 225 cases of biological control programme of about 110 species of pests in 60 countries.

Rao *et. al.* (1971) reviewed the biological control of insects and other pests in South East Asia and the Pacific region. The first dramatic example of deliberate manipulation of insect natural enemies was the importation of the vedalia lady beetle, *Rodolia cardinalis* (Mulsant) in California in 1888 to control the cottony cushion scale, *Icerya purchasi* Maskell on citrus. It is estimated that possibly upto one-third of the successful biological insect pest control programmes are attributable to the introduction and release of insect predators (Sweetman, 1958).

The Commonwealth Institute of Biological Control (CIBC) was established in 1957 at Bangalore and gave tremendous encouragement for Indian Scientists to work on biological control. Since then a number of centres were initiated in our country. Notable among them are the Central Biological Control Station at Gorakhpur (U.P.), Solan (H.P.), Hyderabad (A.P.), Sriganganagar (Rajasthan), Indian Agricultural Research Institute, Pusa (New Delhi), International Crop Research Institute for the Semi Arid Tropics, Patanchery (A.P.), Biocontrol Research Laboratory Chengalpattu (A.P.), Annamalai University, Annamalainagar (T.N.), Marathawada University, Aurangabad, Shivaji University, Kolhapur, G.K.V.K., Bangalore, E.R.I. Madras, etc.

CIBC Bangalore have made extensive survey of important insect pests and their natural enemies in India. Some prominent individuals include aphids, rhinoceros beetles, coconut catterpillar, etc.

In India *C. montrouzieri* was introduced in 1898 from Australia to control *Coccus viridis* (Green) as a very firstly desinged programme for biological control (Puttarudriah *et. al.*, 1952; Mayne, 1953). Later in 1927 lady bird beetles were introduced to control aphids on lentil at Pusa (Dutt, 1927). With the spread of the Pest to Mysore, Coorg, Travancore and Bombay, the citrus crop in this states was seriously threatened. A substation for multiplication and liberation of *Rodolia* was set up at Kodaikanal with stocks obtained from Nilgiris and New Zealand. Over a period of several years many thousands of adults were released in infested localities in small batches and at intervals, resulting in effective control of the pest (Subramanyam, 1955). *C. montrouzieri* has been breed on the Cochineal insect *Dactylopius opuntiae* (Ckll.) in the laboratory and successfully used to control *Pulvinaria maxima* Green on neem, *Centrococcus insolitus* (Green) on brinjal, *P. psidii* Mask. on guava and sapota and *Pseudococcus corymbatus* Green on citrus in South India (Rao and David, 1958). From India *R. cardinalis* and *Rodolia* sp. were sent to Portugal in 1961 and in North West Africa in 1963 for the control of *I. purchasi,* has become established and gave good control (Simmonds, 1971). The Coccinellids *C. montrouzieri, Nephus* sp. and *Hyperaspis* sp. from India were also sent to the Bahamas and Barbados in 1968-69 for the control of soft scales on guava and *Saccharicoccus* sp. (Pink mealy bug) on sugarcane and have established successfully. In general, several predators have shown their tremendous potential for the control of various insect pests in India.

The family Coccinellidae is extremely important from the view of biological control of insect pests. Thousands of predators act on the poopulation densities of certain target pest species, hence they have been utilised extensively in biological control programmes. The predators have been reported on Homopterans like, aphids, psyllids, coccids, mealy bugs and on non-insects like mites. Most of the past moddling work is devoted to their seasonal abundance, appearance records, preys and utilisation in biocontrol, etc. but very little attention is paid on their taxonomy, biology and behaviour.

In past, taxonomical studies of Coccinellids have been attempted by some workers (Linnaeus, 1758; Fabricius, 1775; Mulsant, 1866; Lewis, 1873, 1879; Crotch, 1873; Casey, 1899, 1908; Timberlake, 1943; Brown, 1962; Chapin, 1965; etc.). Korschefsky (1931, 1932, 1933) has described 29 species from USSR. Ghani (1962) reported 10 species with their descriptions from Pakistan.

Kamiya (1966) described 25 species of different tribes of family Coccinellidae from Japan. In addition, Miyatake (1967, 1969, 1970, 1985) also described 53 species with 10 new species from Japan. Similarly Sasaji (1967, 1971) reported 148 species from Japan. A taxonomic studies of the Coccinellidae of Louisiana was undertaken in 1965 by Chapin (1974) to identify the species wherein descriptions are provided for 6 sub families, 11 tribes, 27 genera and 71 species. Of which the species listed, 38 were reported for the first time and 5 were newly described. Rafiq Ahmed (1973) investigated new tribe Ghaniini of sub family Chilocorinae of family Coccinellidae from Pakistan under which he reported only one species. From India Kapur (1947, 1948a, 1948b, 1950, 1954, 1967, 1969, 1970, 1972) described 62 new species from various parts namely Andaman, Goa, Manipur, etc., Channa Basvana and Putttrudriah (1953, 1955, 1956) described some species from Karnataka. Similarly Ghorpade (1974) also reported one new species of Coccinellid from Karnataka. In addition, Hassena Bhaskar (1992) redescribed 48 species in the form of M.Sc. thesis from Karnataka. Very recently, Omkar and Bind (1993) described *Coccinella septempunctata* Linn., *C. transversalis* (Fab.), *Menochilns sexmaculatus* (Fab.), *Brumoides suturalis* (Fab.), *Micraspis discolor* (Fab.), and *M. cardoni* (Weise) from Central Uttar Pradesh. Sathe and Bhosale (1994) described one new species of the genus *Coccinella* from Maharastra, India. This work will be useful in pest management by biological means.

2

Collection, Preservation and Rearing

The design of facilities for culturing insects is, or should be, a key topic in the broader subject of insect rearing. Facilities design however, has too often been relegated, figuratively and literally, to the closet and back shelf. The basic requirements for successful rearing are (1) Food, (2) Protection from enemies, (3) a suitable physical environment and (4) Fit conditions for reproduction. Insect rearing is an on going process and more efficient methods are continuously been adopted, sometimes very minor changes in techniques can have drastic results, both positive and negative (Patana, 1975). The application of clear room techniques and best available environment control system greatly facilitate the production of insects rearing in continuous and closed cultures. In the present study various methods have been adopted for rearing the insects, details of them mentioned are as below :

Materials

(1) Glass Cages (Plate-1, Figs. 1 & 2)

For rearing of Coccinellid beetles and prey species, two types of glass cages (Figs. 1 & 2) were used. Both cages are quadrangular (size 25 × 25 × 30 cm) in shape. Each type consists wooden base and glass walls on three sides. In one cage (Fig. 1) 4th side is provided with muslin sleeve for handling the insects and in another case (Fig. 2) the 4th side of the cage consists of a glass window. Both types of cages were used for rearing the predator species and the prey species. To facilitate the rearing technique effectively a twig of a host plant along with preys and predators were fixed in a circular paper disc over the beaker containing water. The cages were also used to observe some behavioural aspects of predators like mating.

(2) Rearing Unit I (Plate-1, Fig. 3)

Rearing Unit I contains a glass chimmney which measures 23cm in height, 4.5cm in diameter at one end, 6.5cm at the another and 9.5cm in diameter at the broadest region, the conical flask (500ml) which measures 17cm in height and a circular paper disc with central hole. For the rearing of predators, conical flask was filled with water and a twig of a host plant along with prey and predator was fixed through the hole of circular paper disc so as to reach the basal portion of the twig into the water. Later, the tiwg was covered with glass chimmney keeping narrow end over the conical flask, the another end of the chimmney was closed by muslin cloth.

(3) Rearing Unit II (Plate-1, Fig. 4)

This unit contains plastic container of size 4.5cm in diameter and 6.5cm in height, empty electric glass bulb and a circular paper disc. A host plant twig was fitted to the paper disc and placed inside the empty glass bulb as shown in figure. The twig was covered by bulb kept over the plastic container so as to reach the plant twig into the water in the container. For the aeration numerous holes were prepared on the upper part of the container. Rearing Unit II is significant from the view point of maintenance of the moisture for the rearing medium.

(4) Glass Trough (Plate-1, Fig. 5)

Glass troughs of 11cm height and 19cm diameter were used for rearing of the Coccinellid beetles. During the rearing the troughs were covered by muslin cloth.

(5) Petridishes (Plate-1, Fig. 6)

Two types of petridishes of 12cm and 10cm in diameter and

PLATE 1

2.2 cm and 2cm in height were used for rearing the predator and prey species and for observing behavioural aspects like mating.

(6) Plastic Containers (Plate-1, Fig. 7 & 8)

Four types of plastic container with 5.5cm, 4.5cm, 4.0cm and 3.2cm in diameter and 7.5cm, 5.0cm, 4.5cm and 3.7cm in height respectively were used for handling and rearing the predator/prey species.

(7) Test Tubes

For handling the predators test tubes (size 19 × 2.5, 15×2, 14.5×2.8cm) were used.

(8) Specimen Tubes (Place-1, Fig. 8)

Specimen tubes of size 1.7cm in diameter and 5.2cm in height were used for handling the insects and preserving various stages of life cycle of beetles.

(9) Insect Boxes (Plate-1, Fig. 9)

Insect boxes (Vaman D. Purohit made) size 44 × 29 × 7.5cm were used for keeping the preserved insects.

Rearing of Predator Species

From the fields of Sorghum, *Sorghum vulgare* Pers. adults, pupae and larvae of both species, *Illeis darbarii* sp. nov. and *Menochilus sexmaculatus* (Fab.) were collected and reared in rearing unit I (Plate-1, Fig. 3) and rearing unit II (Plate-1, Fig. 4) for initial culture. Newly emerged adults from the field collected material, males and females allowed to mate in petridish/glass cages with pairs. Later, the mated females were introduced in the rearing unit I and II for egg laying. The eggs were laid in clusters of 150 to 200 and they hatch within 4 days. Newly emerged larvae were fed with aphids, *Rhopalosiphum maidis* (Fitch) and allow to grow into four instars, pupae and adults. During the rearing of the predator species, their food material (aphids) were changed by 72 hr interval. The aphids were provided to the predators by taking them on the host plant twig as described in rearing unit I and II.

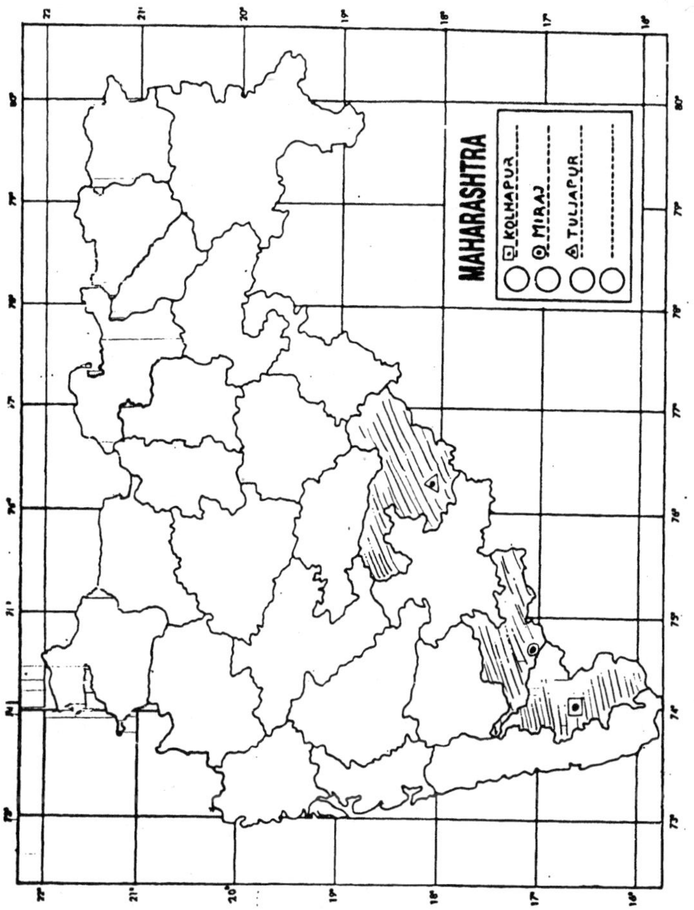

Fig. 10 : Map of Maharashtra

COLLECTION AND PRESERVATION

Survey of coccinellid predators was carried out from Western Maharashtra during 1992-94. A very large number of specimens were collected from agricultural crops *viz.*, Jowar, *Sorghum vulgare* Pers., Maize, *Zea mays* Linn.; Groundnut, *Arachis hypogaea* Linn.; Soyabean, *Glycine soja* Linn.; Cow pea, *Vigna catjang* Walp.; Sunflower, *Helianthus annus* Linn.; Safflower, *Carthumus tinctorius* Linn., etc. and from non agricultural, Subabul, *Leucaena retusa*, Brew.; Rui, *Calotropis gignatea* Linn.; China rose, *Hibiscus rosasinensis* and composite garden plant, *Gaillardia bulchella* Fonger. The concinellid species described in this thesis were also collected from ecologically varying types of habitat, like agricultural fields, orchards, forests, etc.

The most beneficial time for collecting lady beetles was from early spring to mid summer, but collections were made throughout the year. Specimens usually collected by sweeping or beating vegetation in cultivated and wooded areas and by hand picking.

Specimens were killed and then pinned and preserved in 70 per cent alcohol. Specimens preserved in alcohol were dissected to study the taxonomic characters. However, colours sometimes faded or specimens with delicate integuments became somewhat distorted after drying.

To study the genitalia, specimens were dissected with the help of dissecting needle then genitalia along with associated abdominal parts were placed in hot solution of concentrated aqueous KOH for about 3 to 5 min. (The prolonged exposure to KOH causes distortion of the genitalia.) Genitalia were then mounted in DPX.

Many times larvae of predatory conccinellidae were collected from field and reared in the laboratory for their adult emergence. Collections were made early in the morning and evening throughout the year. After sorting of different groups and genera, each field collection was duly labelled with date of collection, locality and name of collector. Then antennae, labrum, mandible, maxilla, labium and hind leg were mounted in DPX on slides.

Morphological studies were carried out with the help of monocular microscope. Measurements were noted with eyepiece

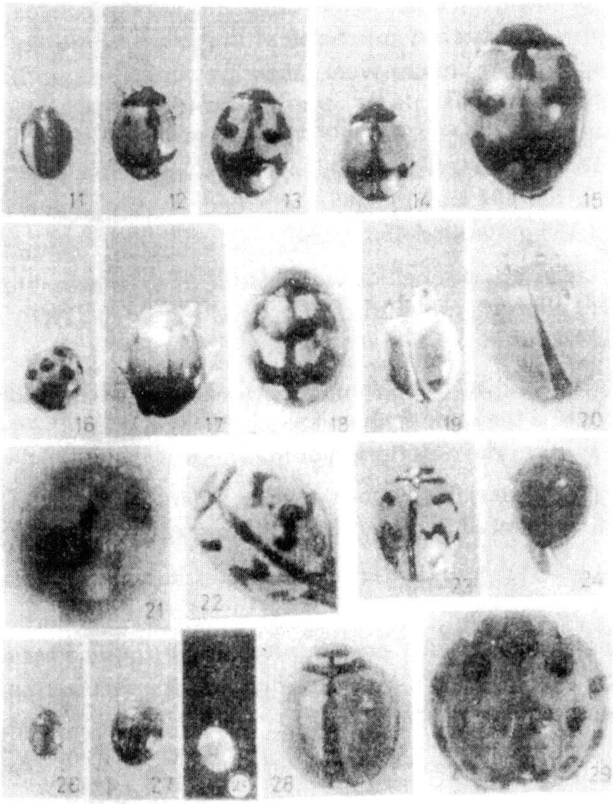

PLATE - 2

Fig. 11 : *Brumoides kolhapurensis* (female) sp. nov.
Fig. 12 : *Coccinella madhuri* (female) sp. nov.
Fig. 13 : *Coccinella transversalis* [Fab.] (female}
Fig. 14 : *Coccinella bhuimungi* (female) sp. nov.
Fig. 15 : *Coccinella yaminii* (female) sp. nov.
Fig. 16 : *Coelophora orientalis* (male) sp. nov.
Fig. 17 : *Harmonia soyabinii* (female) sp. nov.
Fig. 18 : *Horniolus mirajensis* (male) sp. nov.
Fig. 19 : *Illeis sathei* (male) sp. nov.
Fig. 20 : *Illeis darbarii* (female) sp. nov.
Fig. 21 : *Menochilus tuljapurensis* (female) sp. nov.
Fig. 22 : *Menochilus sexmaculatus* (Fab.) (female)
Fig. 23 : *Menochilus shivajensis* (female) sp. nov.
Fig. 24 : *Menochilus kharipi* (female) sp. nov.
Fig. 25 : *Scymnus (Scymnus) indica* (female) sp. nov.
Fig. 26 : *Chilocorus suryaphuli* (male) sp. nov.
Fig. 27 : *Scymnus (Pullus) marathi* (female) sp. nov.
Fig. 28 : *Vernia polyphagi* (male) sp. nov.
Fig. 29 : *Coccinella yoginii* (female) sp. nov.

micrometer and are in mm. All measurements were calculated with the help of graduated mechanical stage micrometer. The photographs of specimens were taken by using close up lenses while taxonomical parts were photographed microphotographically (Figs. 63-67, 69, 70, 72-79, 80-85, 87, 96-101, 103, 104 and 183-186 magnified 50 times; Figs. 40-48, 52-54, 68, 71, 86, 88-95, 102, 105-154 and 156-165 magnified 100 times; and Figs. 49-51 and 155 magnified 450 times).

The collected coccinellids were identified to be belonging to genera of different subfamilies proposed by Sasaji (1968) and Kapur (1963).

To facilitate exact understanding of the terms, the terminology adopted here is the same as that of Sasaji (1968), Kapur (1970) and Chapin (1974) in the description of the species.

The type material is for the time being in the collection of T.V. Sathe, Department of Zoology, Shivaji University, Kolhapur.

A large number of references were consulted in the course of the studies, and cited in the text of the thesis.

The following terms are adopted in the thesis for head, antennae, mandible, maxilla, labium, hind leg, spermatheca sipho, tegmen and male genitalia.

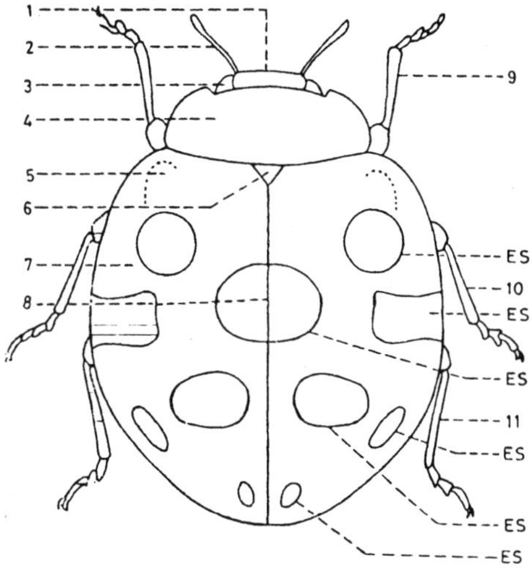

Fig. 30 : Adult dorsal view
(1) Head; (2) Antenna; (3) Eye; (4) Pronotum; (5) Humeral callcus; (6) Scutellum; (7) Elytra; (8) Elytral suture; (9) Fore leg; (10) Middle leg; (11) Hind leg; ES = Elytral spots

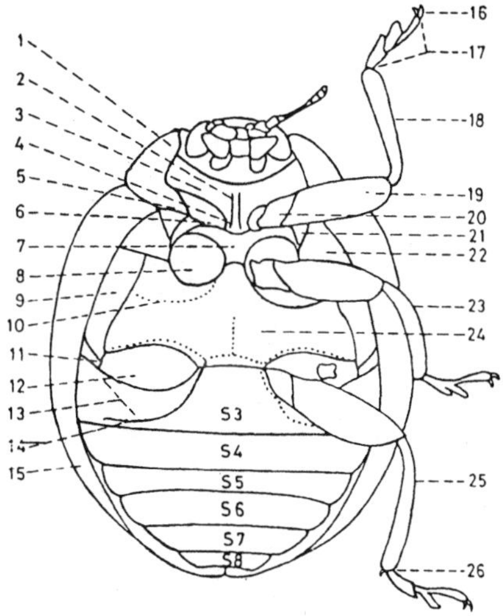

Fig. 31 : Adult ventral view
(1) Hypomeron; (2) Prosternal carina; (3) Prosternum; (4) Prosternal process; (5) Postcoxal process; (6) Fore coxal cavity; (7) Mesosternum; (8) Middle coxal cavity; (9) Metepisternum; (10) Femoral line; (11) Metepimeron; (12) Hind coxal cavity; (13) Lateral femoral line; (14) Femoral line; (15) Elytral epipleuron; (16) Claw; (17) Tarsus; (18) Tibia; (19) Femur; (20) Trochanter; (21) Mesepisternum; (22) Mesepimeron; (23) Tibia of middle leg; (24) Metasternum; (25) Tibia of hind leg; (26) Tibial spur of hind leg; S3-S8 = Sternite of abdominal segments

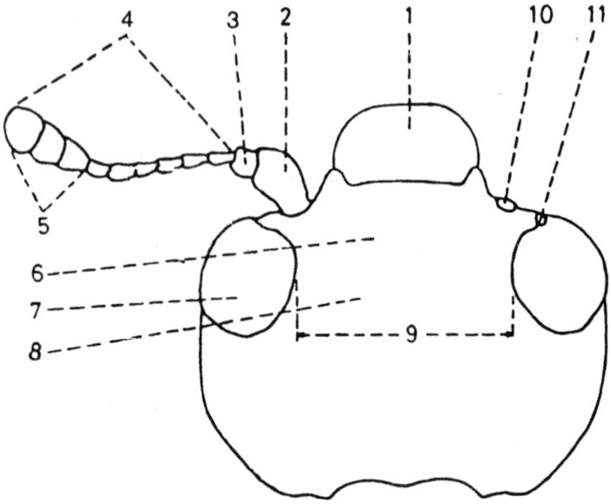

Fig. 32 : Head dorsal view
(1) Labrum; (2) Scape of antenna; (3) Pedicel of antenna; (4) Flagellum of antenna; (5) Antennal club; (6) Clypeus; (7) Compound eye; (8) Frons; (9) Interocular distance; (10) Antennal socket; (11) Post antennal process

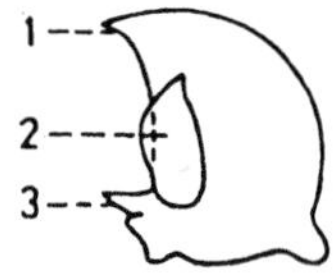

Fig. 33 : Mandible
(1) Apical tooth; (2) Membranous prostheca; (3) Basal tooth

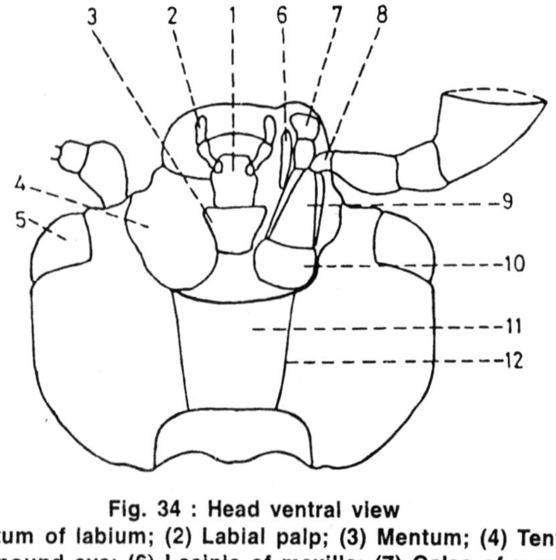

Fig. 34 : Head ventral view
(1) Prementum of labium; (2) Labial palp; (3) Mentum; (4) Tentorium; (5) Compound eye; (6) Lacinia of maxilla; (7) Galea of maxilla; (8) Maxillary palp; (9) Stipe of maxilla; (10) Cardo maxilla; (11) Gula; (12) Gular suture

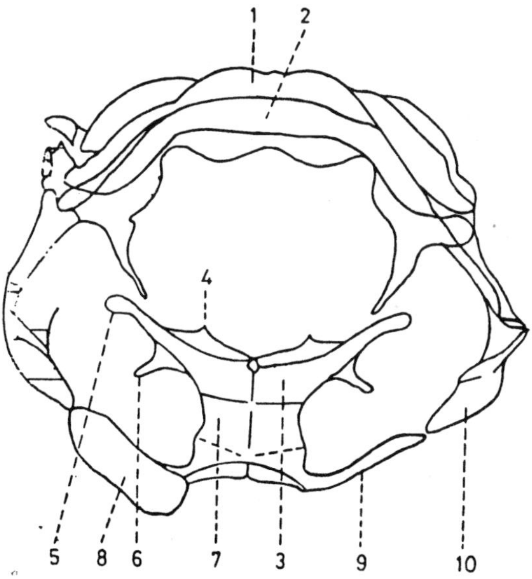

Fig. 35 : Metathorax
(1) Metathorax; (2) Postnotum of metathorax; (3) Metendosternite; (4) Anterior tendon of metendosternite; (5) Lateral arm of metendosternite; (6) Lateral lobe of metendosternite; (7) Sternum; (8) Coxa of hind leg; (9) Metasternum; (10) Metepimeron

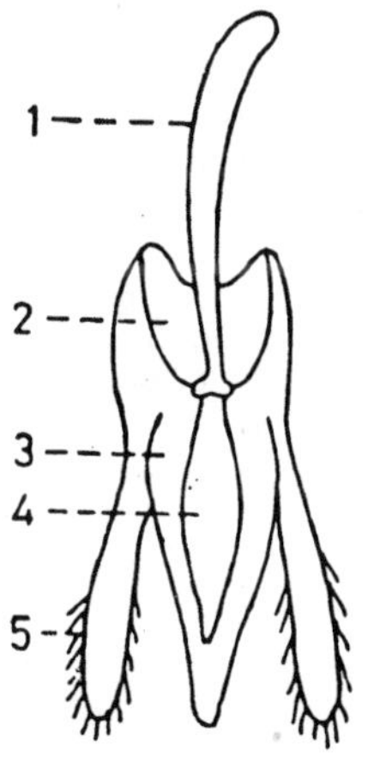

Fig. 36 : Tegmen of male
(1) Tegmen strut; (2) Basal piece; (3) Median lobe; (4) Groove; (5) Lateral lobe

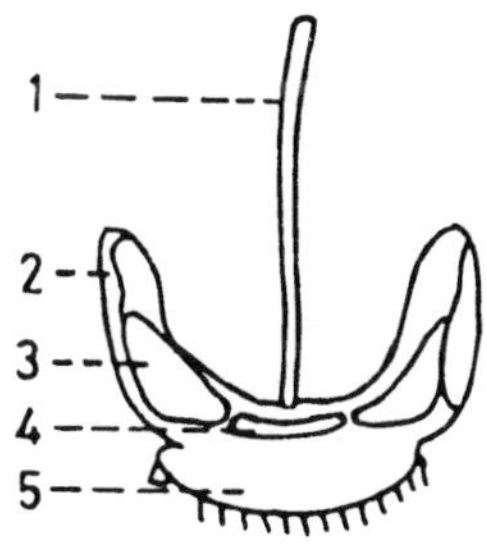

Fig. 37 : Male genitalia
(1) Apophysis; (2) 9th tergite; (3) 9th pleurite; (4) 9th sternite; (5) 10th tegite

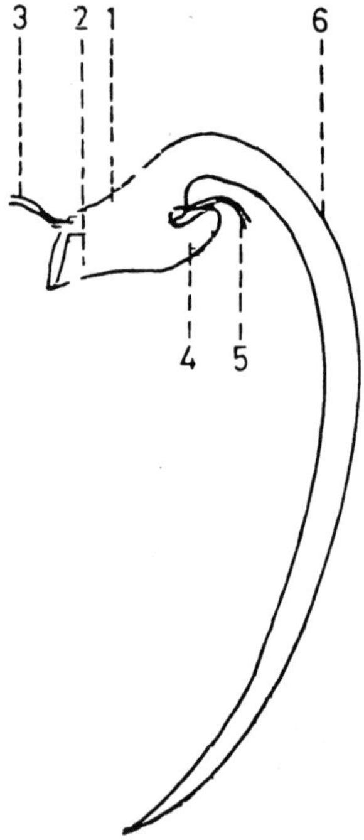

Fig. 38 : Sipho
(1) Siphonal capsule; (2) Outer lobe of siphonal capsule; (3) Ejaculatory duct; (4) Inner lobe of siphonal capsule; (5) Trachea; (6) Siphonal tube

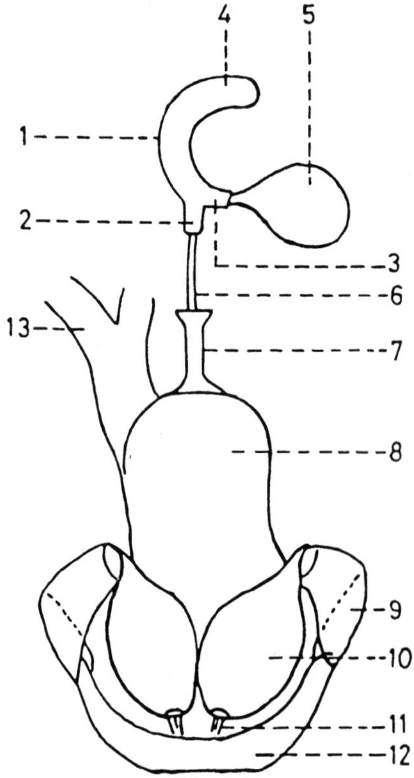

Fig. 39 : Ovipositor and spermatheca
(1) Receptaculum seminis; (2) Nodulus; (3) Ramus; (4) Cornu; (5) Accessory gland; (6) Sperm duct; (7) Infundibulum; (8) Bursa copulatrix; (9) 9th pleurite; (10) Hemisternite; (11) Stylus; (12) 9th tergite; (13) Oviduct

3

TAXONOMY

Taxonomy has never been an attractive profession. Firstly, it is regarded as unrewarding; secondly it is given step-motherly treatment from financial agencies and thirdly the students find field like molecular biology. Presently, the picture is very different and fast changing in favour of taxonomy. Taxonomists are presently employed by Universities, Research Institutes, Museums, Central and Government agencies, Industries, Zoos, etc. Taxonomy plays a quite wider role in the important fields of applied biology including public health, environmental problems, national defence, wild life management, pest management in identifying the species. The systematists are presently greatly involved in designing and implementing biological control programmes of pests and diseases most effectively. The knowledge of an organism's relationships with the other permit us to develop means of control more easily. Total dependant on correct identification of both biocontrol agents and target pests is not ruled out. Recently, increasing efforts have been directed towards the biological control programme of pests, since biological control is very effective alternative for pesticidal control. Biological control and taxonomy are inter-related and interdependent branches of science. Taxonomist needs for identification of the species, to study the evolutionary history and guide the explorations for native and exotic species of biocontrol agent.

In the biological control programme coccinellids have been used extensively. The coccinellid beetles are familiar to people because of their prettiness and scientific importance. Out of estimated, more than 3,20,000 coleopteran species in the world,

4,780 have been described from India (Lefroy 1909). Among the various families of coleoptera, coccinellidae has outstanding importance since most of the species are predaceous on other insect pests and hence widely used in biological pest suppression. Coccinellids feeds on larval and adult stages of aphids, scale insects, psyllids, phytophagous mites, etc., which are injurious to various agricultural and forest crops. The coccinellids have also great importance from the viewpoint of evolutionary biology, population ecology, genetics, cytology, biogeography, etc.

About 490 genera and 4200 species of coccinellidae are known in the world (Sasaji, 1971). From Japan, 49 genera and 153 species have been reported.

Linnaeus (1758) first described 36 species under the genus *Coccinella*. Later, Redtenbacher (1843) worked on the sub division of the family Coccinellidae into aphidophagous and phytophagous groups. Mulsant pubished a monograph "Securipalpus" on coccinellids. Later, in 1850, Mulsant established his system for all the coccinellids of the world. In his system he divided coccinellidae into "Gymnosomides" and "Trichosomides" based on presence or absence of dorsal pubescene. In 1874, Crotch throughly revised the Mulsant's classification and established his own system. He combined *Platynaspis* with *chilocorous* and its allies and he treated epilachnides and coccinellids to be belonged to a single sub family coccinellidae. But, in 1876, Chapius revised the family coccinellidae into two sub divisions namely "Aphidophagous" and "Phytophagous". Further, Chapius divided the aphidophagous into 13 groups. Later, during 1877 to 1925, Weise published more than one hundred papers on taxonomy of the family Coccinellidae. He revised numerous genera and species and gave the perspective of the systematic position of the genera. In 1885, Ganglbauer revised the family into two major groups, "Coccinellidae aphidiphagae" and "Coccinellidae phytophagae" and the former further into 7 groups. In 1887, he formed a distinct group "Pseudococcinellidae" of the genera *Coelopterus*, *Pharus* and *Sticholotis*. Again, he published a important article on Ceylonese coccinellidae wherein he subdivided the family coccinellidae into Aspidimerini and Clanini. In 1899, Ganglbauer divided the family coccinellidae into 3 sub families. He erected an independent sub family, lithophilinae for *Lithophilus*, based on tetramerous tarsal formula. His system has long been followed by the later authorities. Similarly, in 1899, Casey revised the American Coleoptera, he separated the family coccinellidae into two major parts, Aphidophagous and phytophagous, wherein he arranged 16 tribes but not any sub-families. His arrangement of the tribes was as follows : Hippodamiini, Coccinellini, Psylloborini, Scymnillini, Hyperaspini, Cranophorini, Epilachnini, Pentilini, Chilocorini, Platynaspini, Telsimiini, Pharini, Oeneini, Scymnini, Rhyzobiini and Coccidulini. He has described a very large number of the species from the world including North America. However, as his opinion Epilachnini is not separable from the aphidophagous groups but, it has not been followed by the later authorities.

Sicard (1907, 09), who did numerous investigations on the Coccinellidae, has arranged a long series of Madagascan species. This system was similar to that of Weise, but he separated the genus *Sticholotis* and its allies with a conical apical segment of the maxillary palpus as a different group. In 1924, Dobazhansky separated the tribe Stithorini from the Scymmnini for the single genus *Stethorus*. He (1926) also discussed the phylogenetic relationship among the family based on the female genital organs.

Korschefsky (1931, 32) arranged all the coccinellidae of the world by the following system :

Epilachninae	10 genera,
Lithophilinae	*Lithophilus*,
Coccinellinae	
Coccidulini	23 genera (including Rhizobiini),
Noviini	5 genera,
Ortaliini	17 genera,
Scymnini	20 genera (including Stethorini),
Scymnillini	*Scymnillus*, *Zagloba*,
Aspidimerini	*Aspidimerus*, *Cryptogonus*, *Cyrema*, *Hypocyrema*,
Cranophorini	8 genera,
Hyperaspini	8 genera,
Pharini	20 genera (including Sticholotini, Coleopterini, Telsimiini, Serangiini,
Oeneini	9 genera,
Clanini	*Jauravia*,
Pentiliini	*Pseuodosmilia*, *Pentilia*

Exoplectrini	10 genera (including *Chnoodini*),
Azyini	*Azya, Ladoria,*
Platynaspini	*Platynaspis, Boschalia,*
Chilocorini	15 genera,
Synonychini	30 genera,
Coccinellini	38 genera (including Hippodamini, Anisostictini),
Psylloborini	13 genera and
Discotomini	5 genera,

Korschefsky's system has long been a standard of the classfication, though several trials on the systematic position of each genus have been undertaken. Later, Watson (1956) discussed the phylogeny of Coccinellini. Mader (1926-37, 41, 51, 54, 55) has published two series of large works on the classification of coccinellidae. His system was almost similar to the Korschefsky's one. Some Authors (Arnett 1962, Kapur 1963, Zaslavskij 1965) followed the Koreschefsky's key and revision with slight modification.

In 1968, Sasaji established the following new system from the view point of phylogeny, based on comparative morphology of both larvae and adults :

Family *COCCINELLIDAE*

Sub family – sticholotinae – Sukhumhikonini, Serangiini, Sticholotini, Shrioznellino,

Sub family – scymninae – Stethorini, Scymnini, Aspidimerini, Hyperaspini, Ortaliini,

Sub family – chilocorinae – Telsmiini, Platynaspini, Chilocorini,

Sub family – Coccidulinae – Lithophilini, Coccidulini, Exoplectrini, Noviini,

Sub family – Coccinellinae – Coccinellini, Psylloborini and

Sub family – epilachnini.

Several recent workers (Ahmed 1973, Chapin 1984, Gorden 1989, Bhasker 1992, etc.) followed the system of Sasaji.

Ahmed (1973) described a new tribe Ghanini for inclusion of the genus *Ghanius* Ahmed. The species of this tribe fed on scale insects in Pakistan. Chapin (1974) published a monograph "the Coccinellidae of Louisiana" wherein he described 71 species belonging to 27 genera, 11 tribes and 6 sub families. Gorden *et al.* (1989) erected a new tribe Carinodulini for the inclusion of *Carinodula campbelli* Gorden *et al.* from Mexico.

From India Subramaniam (1923) gave a list of some coccinellids of South India along with their hosts distribution. Aiyer (1924) described a new species, *Scymnus coccivora* preying on the neem scale, *Chloropulvinaria maxima* (Green). Kapur (1948a) studied the old world species of the genus *Stethorus* Weise and descirbed 12 new species. He also revised (1948b) Indian species of *Rodolia* Mulsant and added 3 new species. Puttarudriah and Channabasavanna (1953, 1955, 1956) recorded 53 species in 23 genera grouped in 8 tribes and 5 sub families. Usman and Puttarudriah (1955) listed 48 species of predaceous coccinellidae from the erstwhile Mysore State. Kapur (1954) provided notes on systematics and biology of lady bird beetles predaceous on the San Jose scale, *Quadraspidiotus perniciosus* (Comstock) in Kashmir. Later, Kapur (1961) described a new species *Stethorus keralicus* Kapur feeding on areca palm mites. Kapur and Rao (1962) described *Adalia simmondsi* as a new speices feeding on *Adelges* spp. in North Western India. Nagraja and Hussainy (1968) gave descriptions, notes on distribution and diagnostic key for 6 species of *Chilocorus*. Kapur and Munshi (1970) described a new species, *Scymnus* (*Pullus*) *nymphaeus* predaceous on the aphid, *Rhopalsiphum nymphae* (Linnaeus) in Calcutta. In 1972, Kapur also recorded 17 species of coccinellidae from Goa. Ghorpade (1974) described a new species, *Cryptogonus kapuri* of the tribe Aspidimerini from Bengalore. He also described a new species *Pseudoscymnus dwipakalpa* preying on *Aspidiotus destructor* Signoret from Bangalore (1977). In 1978, Gorden described several West Indian species of coccinellidae, mostly scale predators representing 5 tribes. Pajni and Singh (1982) recorded 30 species of coccinllidae belonging to 18 genera and 2 sub families from Chandigarh. Gorden and Chapin (1983) revised the mite feeding coccinellid genus *Stethorus* in the Western hemisphere recognising 21 species, of which 11 were new. Pajni and Verma (1985) described and illustrated the male genitalia of 25 coccinellid species belonging to 15 genera collected from Chandigarh. Canepari (1986) studied the type material of 36 species of coccinellidae of Northern India and Nepal in Geneva Museum. Booth (1991) described a new genus and species of the tribe Scymnini associated with *Rostrococcus* spp. from Southern India. Haseena Bhasker (1992) redescribed 31 coccinellid spp. in the form of thesis (M.Sc.) from Bangalore. The review of literature shows that very little attention is paid on the coccinellids from Maharashtra. Keeping in view the above fact the topic was selected.

The family coccinellidae is characterised by : Body usually oval and round and more or less strongly convex; antennae more or less clavate; terminal segment of maxillary papli usually englarged and triangular (Securiform), but often parallel sided or conical; elytra not at all truncate and never distinctly straight; hind coxae strong transverse; abdomen with accurate femoral lines; tarsal formula usually cryptotetramerous; tibial spurs simple and minute; tarsal calws simple or apendiculate; sipho of the male genitalia always elongate and more or less curved ventrally; tegmen trilobed.

The family coccinellidae contains 6 sub families *viz.*, sticholotinae, scymninae, chilocorinae, coccidulinae, coccinellinae and epilachninae. Among these scymninae, chilocorinae and coccinellinae are important since they frequently seen on homopterous insect pests.

Sub-family - Coccinellinae

The sub family represented by following characters :

1. Body medium to large sized,
2. Dorsal surface glabrous,
3. Antennae long, 11 segmented and inserted more or less dorsally,
4. Mandible with basal tooth, bifid appically,
5. Apical segment of maxilllary palpi distinctly securiform,
6. Abdomen always composed of 6 visible sterna,
7. Femora elongate and not flattened and

8. Tarsi always cryptotetramerous.

The sub family coccinellinae contains two tribes *viz.*, Coccinellini Weise, 1985 and Psylloborni Casey, 1899.

Tribe Coccinellini Weise, *1885.*

Coccinellids Leach, 1815, *Brewstar, Edingh, Encyclop.*, 116.

Coccinellids Mulsant, 1846, *Securipalp* : 28.

Coccinellids Weise, 1885, *Best, Tab, europ. Coleopt.* 11, ed. 2 : 2.

The tribe coccinellini easily distinguishable from other by following characters :

1. Body large to medium sized, usually short oval or hemispherical, rarely elongate oval,
2. Dorsum always glabrous,
3. Antennae at least nearly as long as interocular distance, eleven segmented and weakly but distinctly clavate,
4. Apical segment of maxillary palpus usually securiform and
5. Mandible with a bifid tip and a basal tooth.

This tribe includes 22 genera.

Tribe Psylloborini Casey *1899.*

Psylloborini Casey, 1899, *J.N.Y. ent. Soc.* 7:93, 100.

The tribe Psylloborini shows following characters :

1. Mandible usually with multidenticulted apex, rarely with a bifid apex,
2. Anterior margin of pronotum very weakly and accurately concave,
3. Galea of maxilla very broad and
4. Usually mycetophagous.

The tribe psylloborini is very much closely related to the tribe coccinellini. It includes 3 genera namely, *Vibidia* Mulsant, 1846; *Illeis* Mulsant, 1850 and *Halyzia* Mulsant, 1850.

Sub-family - Chilocorinae

The sub family is clearly defined from the other sub families of the coccinellidae by following characters :

1. Clypeus distinctly broadly and laterally expanded,
2. Antennae usually reduced in the length and the segments,
3. Terminal segment of maxillary palpi never strongly divergent apically and
4. Elytral base distinctly broader than pronotal base.

Subfamily Chilocorinae is divided into three tribes, Chilocorini Mulsant, 1846; Platynaspini Mulsant, 1846 and Telsimiini Casey, 1899.

Tribe Chilocorini Mulsant, *1846*.

Chilocorines Mulsant, 1846, *Securipalp* : 166.

This tribe shares for the sub family Chilocorinae by its major number. The tribe is visualised by,

1. Dorsum glabrous, rarely pubescent,
2. Femora normal, not strongly depressed,
3. Tarsi cryptotetramerous and
4. Abdomen composed of 5 visible segments in female and 6 in male.

The tribe chilocorini includes 2 genera, *Chilocorus* Leach, 1815 and *Arwana* Leng, 1908.

Sub-family – Scymninae

The Sub family is represented by following characters:

1. Body usually small to medium sized,
2. Antennae usually very short,
3. Terminal segment of maxillar palpi usually nearly parallel sided and
4. Elytral base not much broader than pronotal base.

Sasaji included 5 tribes under this sub family namely,

Hyperaspini Mulsant, 1846; Scymnini Costa, 1849; Ortalini Mulsant, 1850; Aspidimerini Weise, 1900 and Stethorini Dobzhansky, 1924.

Tribe – Scymini Costa 1849

Scymniens Mulsant, 1846 *Securipalpes* : 210

Scymnini Costa, 1849 P. 9.

The tribe Scymnini is one of the largest group in the sub family and defined as follows :

1. Body usually less than 4 mm, rarely medium sized,
2. Body shape elongate oval to hemispherical,
3. Clypeus not widely and strongly expanded laterally,
4. Terminal segments of maxillary palpi parallel sided or slightly divergent apically,
5. Anterior margin of prosternum flat or slightly concave and
6. Elytral epipleura very narrow, horizontal and without foveae.

The tribe Scymnini includes 7 genera *viz.*, *Scymnus* Kugelann 1794; *Nephus* Mulsant, 1846; *Cryptolaemus* Mulsant, 1853, *Horniolus* Weise, 1900; *Pseudoscymnus* Chapin, 1962; *Axinoscymnus* Kamiya, 1963 and *Keiscymnus* Sasaji, 1971.

The present chapter deals with the taxonomic details of predaceous coccinellidae from Western Maharashtra, India. This is first attempt on coccinelidae from this region. The study covers the description of the species, distribution, host records and parasitioid records. In all seventeen new species have been described belonging to ten genera of four tribes of three sub families of the family coccinellidae and two species of coccinellids have been redescribed. The present study would be helpful in biocontrol programme of pests, ecology, evolutionary studies, for making prey-predator index and predator-Parasitoid index, etc.

Genus *Brumoides* Chapin, 1965

***Brumoides* Chapin, 1965.**

Bull. Mus. Comp. Zool. Harvard Univ. 133(4) : 237. Type species : *Coccinella suturalis* Fabricius.

The genus *Brumoides* is erected by Chapin in 1965. It is distributed in Oriental, Nearctic or Ethiopian regions. The genus *Brumoides* is separated from the genus *Brumus* Mulsant, 1850 by having 8 segmented antennae. The species assigned to this genus are *B. suturalis* Fab., *B. lineatus* Weise, *B. hainanensis* Miyatake, *B. maai* Miyatake and *B. ohtai* Miyatake. From India, *B. suturalis* is the only species redescribed by Bhaskar (1992) and Omkar and Bind (1993).

The genus is characterised by oval body, moderately convex and glabrous; antenna 8 segmented, 1st segment short and stout; outer margin of mandible strongly convex; lacinia with row of three short, stout spines; labial palp long and slender; median lobe of male genitalia triangular, sipho moderately slender; cornu small, hemisternites moderately long, styli small, button shaped with 2 or 3 long setae.

Brumoides kolhapurensis sp. nov.

(Plate-2, Fig. 11; Plate-6, Figs 40-48)

Female

Body short, elongately oval, 4.00 mm long and 3.26 mm wide, moderately convex on dorsal side, glabrous; pronotum reddish brown, 0.78 mm long and 1.85 mm wide; antenna 8 segmented, 0.44 mm long; elytra 3.31 mm long and 2.02 mm wide with sutural and discal blackish vittae, finely and closely punctate; inner surface of spermatheca "V" shaped, 0.21 mm long with elongated hook like cornu.

Head

Yellowish brown, 1.05 mm long and 1.30 mm wide; completely inserted into pronotum; frons flats, smooth; fronto clypeal suture indistinct; clypeus slightly expanded in front of compound eyes; interocular distance 0.73 mm; gular sutures distinct and runs parallel; tentorium with parallel arms. *Antenna (Fig. 40) : 0.44 mm* long, 8 segmented; scape large as wide long; pedicel as broad as scape at base; 1st segment small; 2nd smaller than 3rd segment; 4th segment large obliquely apically truncate; 5th segment somewhat quadrate; 6th segment small partly embedded in apex of 5th segment with 2 to 3 long setae.

Flagellar formula : SL/W=0.75, PL/W=1, IL/W=0.66, 3L/W=1, 4L/W=1.33, 5L/W=0.5, 6L/W=0.66.

Labrum : Small, 0.29 mm long and 0.44 mm wide, roundly oval, plate like with thick setae on anterior and lateral borders. *Mandible (Fig.41)* : Somewhat triangular, 0.40 mm long and 0.29 wide in middle, outer portion yellowish, inner portion dary brown, outer basal portion concave and strongly convex toward spical side; apical tooth bifid, sharp, pointed with depression 0.21 mm; basal tooth sharp pointed; membranous prostheca 0.18 mm. *Maxilla (Fig.42)* : Cardo small, 0.18 mm long and wide; stipe more longer than width, 0.29 mm long; galea somewhat large, 0.21 mm long; lacinia with 4 spines and long setae, 0.16 mm long; maxillary palpus 4 segmented, 0.63 mm long; basal segment small; 2nd narrow at base and broad at apex; 3rd segment transverse; terminal segment large, securiform with truncate apex. *Labium (Fig. 44)* : Submentum fused gula; mentum large, 0.23 mm long 0.25 on wide, somewhat quadrate, broad in middle, anterolateral sides slightly come outwardly; prementum 0.12 mm long and 0.16 mm wide, rounded, cylindrical; labial palpus 3 segmented, 0.27 mm long, basal segment small; 2nd narrow at base, broad at apex; terminal segment cylindroarcuate and roujdly ended.

Thorax

1.68 mm long and 2.63 mm wide; pronotum reddish brown, glabrous, "C" shaped, 0.78 mm long and 1.85 mm wide with rounded lateral sides; basal side convex and finely punctured; prosternum somewhat coarsely punctured and finely and sparsely pubescent; mesosternum rugosely punctured; metasternum finely and sparsely punctured at sides; metendosternite "Y" shaped with pair of fulcra; prosternal carinae runs parallel and divergent towards ap;ical side; scutellum black. *Elytra (Fig. 11)* : Yellowish, 3,31 mm long and 2.02 mm wide, glabrous, finely and closely puncatae with sutural and discal black vittae; apical portion reddish; outer border rounded and inclined below towards apex; discal vittae not touching to basal and apical part of elytron and terminate roundly; width of discal vittae and yellowish internal part same except in the middle, yellowish part slightly wider than discal vittae, width of yellowish part in middle 0.67 mm, black discal vittae 0.58 mm, discal vittae and internal yellow part gradually narrower apically but does not terminate into hook like structure; sutural vittae not touching to apex of elytron; elytral epipleuron transverse without fovae and not reaching to apex. *Hind leg (Fig.46)* : Yellowish; coxa elongated, 0.78 mm long and 0.23 mm wide; trochanter 0.14 mm long and 0.31 mm wide, sub quadrate with very slight denticulate process; femur 0.99 mm long and 0.37 wide, broad, elongated; tibia 0.90 mm long and 0.16 mm wide elongated with tibial spines on surface and prominent 2 tibial spurs 0.08 mm; tarsi 4 segmented, 0.73 mm long, 1st segment elongated, 0.42 mm, somewhat triangular and covered by setae; 2nd segment 0.27 mm long; 3rd segment very smaller, 0.08 mm; terminal segment elongated, 0.29 mm long and terminate into bifid claw; claws are prominent, 0.10 mm long, sharp with triangular process in middle and without basal tooth.

Abdomen

2.11 mm long and 2.32 mm wide, dorsum black; six sternites visible; ventral side pubescent; lateral side densely covered by setae; 1st sternite and middle area of second with sparsely but distinctly stronger punctation with fine pubescence on the whole; coxal line complete; femoral line roundly angulate (Fig. 45). *Female genitalia* : 9th tergite, pair of 9th pleurite and a pair of hemisternite together form ovipositor; hemisternites elongated, pointed at apex and broad at base with stylus. *Receptaculum seminis (Fig. 43)* : Inner surface "V" shaped, 0.21 mm long and 0.42 mm wide; cornu 0.10 mm elongated hook like; nodulus broad.

Male

3.79 mm long and 3.16 mm wide, similar as female except

PLATE - 6

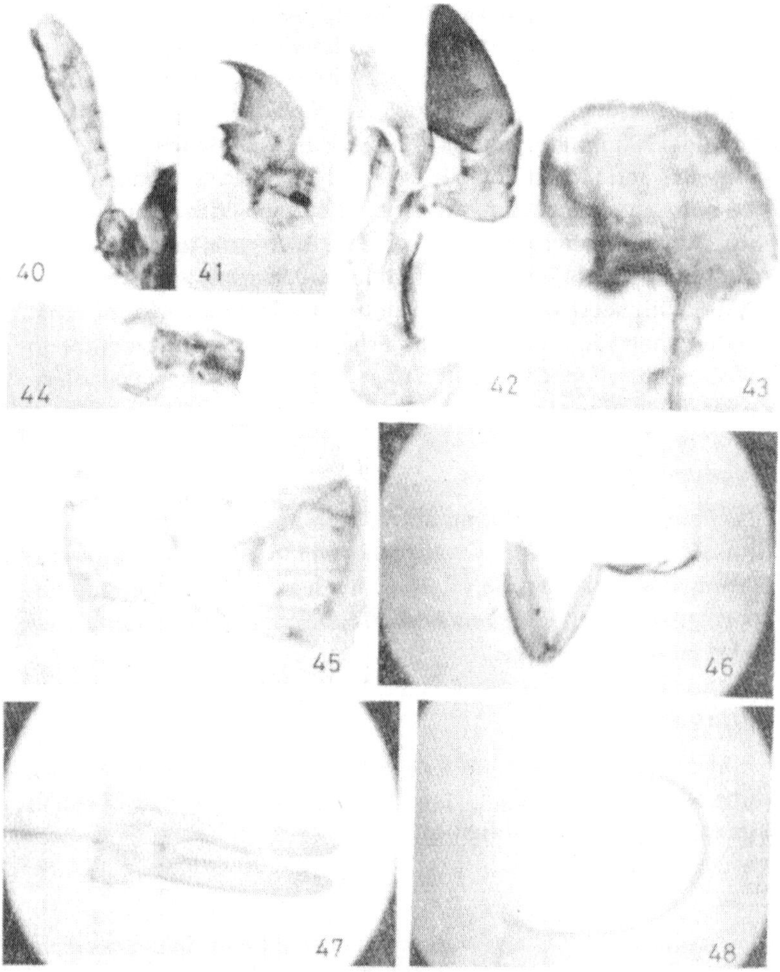

Brumoides kolhapurensis **sp. nov.**

Fig. 40 : Antenna

Fig. 41 : Mandible

Fig. 42 : Maxilla

Fig. 43 : Receptaculum seminis

Fig. 44 : Labium

Fig. 45 : 1st Abdominal sternite

Fig. 46 : Hind leg

Fig. 47 : Tegmen

Fig. 48 : Sipho

body size. *Male genitalia* : 9th tergite somewhat "U" shaped, 0.521 mm long and 1.03 mm wide; 10th tergite strongly transverse with setae; 9th pleurite attached to the 9th tergite; 9th sternite small; apophysis elongated, 0.94 mm long. *Sipho (Fig. 48)* : Siphonal capsule 0.10 mm long and 0.44 mm wide, well developed; outer lobe small and flat, inner lobe elongated and narrow; siphonal tube 3.16 mm long, immediately curved, uniform in width and terminate into swollen structure, apex with swollen part and spatulate structure. *Tegmen (Fig. 47)* : Lateral lobes 0.90 mm long, elongated, robust with setae on apical portion; middle lobe 0.46 mm long, smaller than lateral lobes, triangular with pointed structure to apex; tegmen strut 0.82 mm long, braod at apex; basal piece quadrate, symmetrical.

Holotype

Female, India : Maharashtra, Kolhapur, coll. 28-I-94, Y.A. Bhosale, on Safflower *(Carthamus tinctorius,* Linn.), antenna, labrum, mandible, maxilla, labium, hind leg, 1st abdominal sternite, spermatheca, sipho, tegmen and last abdominal terga of male on slides labelled as above.

Paratype

Female 18, male 29, sex ratio (male : female), 1 : 1.61, data same as in holotype, coll. from 28-I-94 to 10-III-94, Y.A. Bhosale, India : Maharashtra, Kolhapur.

Prey

Aphids, *Dactynotus compositae* Theobald and *Aphis craccivora* Koch.

Prey Plant

Safflower, *C. tinctorius.*

Sunflower, *Helianthus annus* Linn.

DISCUSSION

This species runs close to *B. suturalis, B. lineatus, B. hainanensis, B. maai, B. ohtai* by having 8 segmented antenna, maxilla, labium and number of elytral vittae; it also runs close to *B. ohtai* by having same thoracic structure. But it differes from the above species by having the following characters :

i) Body size large,

ii) Elytral, discal and sutural vittae large and shows the same diameter that of intermediate yellow patches; discal black vittae does not terminate into hook like structure but bluntly ended,

iii) Outer basal portion of mandible is concave and strongly convex towards apical side; inner portion dark brownish,

iv) Femoral line roundly angulate, incomplete and area around femoral line hairy,

v) Hemisternites in female bluntly pointed at apex,

vi) Cornu of spermatheca hook like and narrower,

vii) Lateral lobes of tegmen robust and

viii) Flagellar formula.

Genus *Chilocorus* Leach, 1815

Chilocorus Leach, 1815.

Brewster, Ediobur. Encycl. 9:116.

Type species : *Coccinella cacti* Linnaeus, 1767.

The genus *Chilocorus* is errected by Leach in 1815. It includes the following species : *C. circumdatus* (Gyllenhal, 1808); *c. rubidus* (Hope, 1831); *C. rufitarsis* (Motschulsky, 1853); *C. bijuous* (Mulsant, 1853); *C. chalybeatus* (Gorham, 1892); *C. mikado* (Lewis, 1896); *C. kuwanae* (Silvestri, 1909); *C. takara* (Nakane, 1959); *C. esakii, C. amamensis* and *C. ishioakensis* (Kamiya, 1959); *C. shirozui* and *C. alishanus* (Sasaji, 1968). Three new species have been described by Miyatake in 1969 under this genus (*C. chinensis, C. hupehanus* and *C. gressitti*).

`The genus is characterised by following characters : Drosum glabrous, rarely pubescent; frons flat; antenna 11 segmented, clavate, longer than interocular distance; mandible with bifid apical tooth; cordo of maxilla not strongly expanded laterally; femur normal, slightly depressed, tibial spurs absent, front tibia with a triangular tooth at external side; abdomen composed of five visible segments in female and six in male; femoral line of first abdominal sternum incomplete.

Chilochorus suryaphuli sp. nov

(Plate-2, Fig. 26; Plate-7, Figs, 49-54)

Male

Body small, oval, strongly convex on dorsal side, ventrum black, 3,37 mm long and 2.53 mm wide; pronotum 0.68 mm long and 1.89 mm wide, yellow with lines on dorsal side; antenna 0.90 mm long; elytra yellowish, 2.70 mm long and 1.26 mm wide, finely pubescent with 3 black spots.

Head

Somewhat quadrate, 0.81 mm long and 1.11 mm wide, lateral sides rounded and completely inserted into pronotum; frons flat, yellowish red; clypeus slightly projected anteriorly; eyes projected to the antero-lateral side, surface smooth, interocular distance 0.63 mm; posterior borer covered by hairs; gular sutures indistinct. *Antenna (Fig. 49) :* 11 segmented, clavate, 0.90 mm long, longer than interocular distance; scape large, barrel shapped, longer than width with 2 setae on inner side; pedicel smaller than scape, slightly longer than width with single seta on inner side; 1st segment longer than width with single seta on apex; 2nd to 6th segments more or less equal, without setae; last 3 segments some what larger than other segments, forms club, 9th segment more elongated, apically truncate and apex covered by long and fine setae.

Flagerllar formula : SL/W=1.4, PL/W=1.33, 1L/W=4, 2L/W=1.5, 7L/W=1, 8L/w=1, 9L/W=1.5.

Labrum : 0.14 mm long and 0.19 mm wide, small plate like, lateral sides rounded with setae. *Mandible (Fig. 50) :* small, sub triangular, 0.21 mm long and 0.14 mm wide at base; apical tooth bifid with depression 0.08 mm; basal tooth absent. *Maxilla (Fig. 52) :* cardo 0.15 mm long and wide, as broad as stipe; stipe 0.18 mm long; galea spatulate, 0.12 mm long; lacinia as long as galea and covered by fine setae; maxillary palpus 4 segmented, basal segment small,

0.04 mm long; 2nd slightly longer than 3rd, 0.10 mm long; terminal segment longert than width and obliquely truncate with pointed apex. *Labium (Fig. 51)* ; post mentum fused with gula; mentum 0.09 mm long, triangular; prementum sub quadrate, 0.10 mm long and wide with hairs; labial palpus 3 segmented, 0.14 mm long, terminal segment pointed.

Thorax

1.03 mm long and 2.50 mm wide, pronotum "C" shaped, 0.68 mm long and 1.89 mm wide, yellow with lines on dorsal side, anterior margin clearly marginated, finely punctate, lateral portion narrowly shagreaned; metasternum more depressed and without punctures. *Elytra (Fig. 26)* : 2.70 mm long and 1.26 mm wide, yellowish, finely pubescent and punctate, outer margin black, each elytron with 3 black spots; spots sub spherical, small, median spot touch to inner border of elytron; elytral epipleuron yellowish, slightly transverse and reach to apex. *Hind leg (Fig. 54)* ; coxa small and stout, 0.08 mm long; trochanter somewhat quadrate with slight denticulate process; femur small, broad, 0.11 mm long; tibia 0.13 mm long with tibial spines and with less prominant tibial spurs; tarsi 4 segmented; basal segment small, triangular, 2nd elongated, 3rd very small, quadrate, terminal segment short elongated with broad apex; claw bifid with basal quadrate tooth, 0.05 mm.

Abdomen

1.68 mm long and 2.45 mm wide, six sternites visible, dorsum black, glabrous and smooth; ventrum black, shiny covered by hairs; lateral borders covered by setae; distal end densely covered by some what long setae. Coxal line complete; femoral line angulate and complete, area around femoral line coarsly punctate (*Fig. 53*). *Male genitalia* : 10th tergite transverse covered by setae. *Sipho* : Siphonal capusle well developed, inner lobe of capusle elongated narrowed and outer lobe quadrately small, siphonal tube curved, apex of tube distinctly and strongly sinuated. *Tegmen* : small and stout, median lobe 0.09 mm long slightly smaller than lateral lobes; lateral 0.12 mm long gradually broad towards apex and covered by setae.

PLATE - 7

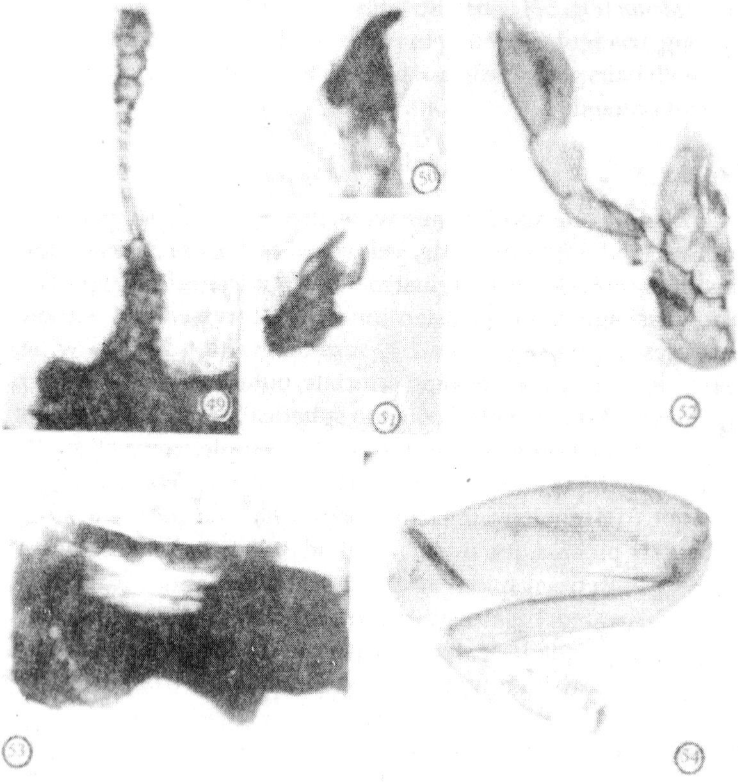

Chilocorus suryaphuli sp. nov.

Fig. 49 : Anteena

Fig. 50 : Mandible

Fig. 51 : Labium

Fig. 52 : Maxilla

Fig. 53 : 1st Abdominal sternite

Fig. 54 : Hind leg

Female

Unknown.

Holotype

Male, India : Maharashtra, Miraj coll., 8-VIII-93, Y.A. Bhosale, on sunflower (*Helianthus annus* Linn.) antenna, labrum, mandible, labium, 1st abdominal sternite on slides labelled as above.

Paratype

Male 3, India ; Maharashtra, Miraj coll. from 8-VIII-93 to 2-XII-93, Y.A. Bhosale.

Prey

Aphid, *Dactylonotus compositae* Theobald.

Prey Plant

Sunflower, *H. annus.*

DISCUSSION

This species runs close to *C. rubidus, C. mikado* and *C. kuwanae* by having body size, antenna, labrum, mandible, sipho and tegmen. However, it differs from the above species by having following characters :

i) Pronotum yellow with lines on dorsal side,

ii) Antenna more longer than interocular distance,

iii) Elytra finely pubescent with 3 black spots on each elytron,

iv) Femoral line on 1st abdominal sternite complete, angulate and area around femoral line coarsely punctured,

v) Siphonal tube distouchly and strongly sinuated and

vi) Flageller formula.

Genus *coccinella* Linnaeus, 1758

The genus *Coccinella* is erected by Linnaeus in 1758 and was primarily holarctic (Dobzansky, 1931). Casey (1899, 1908) stuided the North American species *Coccinella* who recognized more than twice as many forms as separate species as did Leng (1903). Dobzhansky (1931) reported 11 species and many sub species using male genitalia as the principal character. Brown (1962) arranged 18 forms considered to merit names as 12 species. The species described by other workers under this genus are *C. novemnotata* (Herbst, 1793); *C. transversalis* (Fabricius, 1781); *C. nigrita, C. septempunctata, C. repanda* (Thumberg, 1781); *C. dimidia* (Hope, 1831); *C. basalis* (Redtenbacher, 1848); *C. ainu* (Lewis, 1896); *C. explanata, C. haseqawao* (Miyataka, 1963); *C. quinouepunctata* (Linnaeus), *C. transversoquttata quinquenotata* (Felderman), *C. trifasciata perflexa* (Mulsant) and *C. transversoquttat richardsoni* (Brown) from the Peach orchards of Canada. From Ukraine Yadechke (1954) reported *C. septempunctata* to be the typical species of cereal fields. From India, Kapur (1963) reported 3 new species, *C. tibetina, C. lama* and *C. nigroivttata,* he also redescribed the species *C. transversalis* (Fabricius). Diraviam, (1990) observed *C. septempuncatata* on subabul preying on the psyllids.

The genus is characterized by having body broadly oval, convex; head black with two pale spots on frons; antenna slightly longer than frons; post antennal process narrow and long; mesostermum flat to the anterior margin; suctellum triangular; elytra yellow, orange or red with variable black pattern; post coxal line incomplete; middle tibia with spurs.

Coccinella transversalis (Fabricius)

(Plate-2, Fig. 13; Plate-8, Fig. 55-62)

Female

Body large, 5.75 mm long and 4.41 mm wide, oval, strongly convex on dorsal side; pronotum black, 1.68 mm long glabrous; elytra yellowish brown 4.74 mm long with three and half black spots; ventral side black; antenna clavate 1.37 mm long brown with dark black colouration on scape and last three segments; spermatheca 0.40 mm long, "U" shaped.

Head

Black, 1.89 mm long and 2.04 mm wide, quadrilateral with rounded sides; lateral sides covered by setae; frons flat, weakly convex and rough, frontoclypeal suture not distinct; clypeus narrower, broadly expanded laterally and projected anteriorly; compound eyes placed antero-laterally, interocular distance 1.10 mm, surface rough and shiny; gular sutures indistinct, tentorium with pair of very thin parallel arms. *Antenna (Fig. 55)* ; 11 segmented, 1.37 mm long; scape large, as wide as long with 2 long fine setae on lateral sides; pedicel smaller than scape, somewhat quadrilateral, as wide as long with 2 long fine setae on one side; 1st segment longer than other segments of flagellum; last 3 segment large with 3 to 4 long fine setae; 7th and 8th segment with single seta, 9th segment with 4 setae.

Flagellar formula : SL/W=1, PL/W=1, 1L/W=1.75, 5L/W=1, 7L/w=0.57, 8L/W=0.55, 9L/W=0.76.

Labrum (Fig. 58) : somewhat quadrate, 0.42 mm long and 0.90 mm wide, lateral borders rounded and in middle angulate, anterior border covered by long setae. *Mandible (Fig. 56)* : somewhat subtriangular, 0.63 mm long and 0.52 mm wide; apical tooth bifid with 0.20 mm depression; basal tooth pointed, curved and triangular; membranous prostheca 0.25 mm. *Maxilla (Fig. 57)* ; cardo 0.25 mm long and 0.27 mm wide, as wide as stipe; stipe longer than wide, 0.46 mm long and 0.27 mm wide; galea 0.33 mm long, covered by thick, long setae; lacinia 0.33 mm long and covered by thick long setae; maxillary palpus 4 segmented, 1.39 mm long, 1st segment smaller than other, 0.12 mm long, terminal segment securiform, 0.71 mm long and 0.65 mm wide, apically truncate. *Labium (Fig. 59)* ; postmentum indistinct from gula; mentum quadrilateral, 0.21 mm long and 0.35 mm wide; prementum subquadrate, 0.25 mm long and 0.37 mm wide, with fine hairs on outer border; labial palpus 3 segmented, 0.44 mm long; 1st segment 0.10 mm long, small; 2nd 0.14 mm long; 3rd 0.18 mm long, well developed and terminate roundly.

Thorax

1.51 mm long and 3.79 mm wide; prothorax broader than long convergent and inclined anteriorly, anterior margin broadly excavated and cover the head; pronotus black, 1.68 mm long and 3.16 mm wide, glabrous with yellowish spots on anterior corners, anterior corners angulate, posterior border covered by setate; middle coxal cavity narrowly opened outwardly; hind coxal cavity strongly transverse oval, metathorax dorsally entirely concealed by elytra; metendosternite "Y" shaped with broad stalk and paired lateral arms of fulcra; scutellum black. *Elytra (Fig. 13)* ; 4.74 mm long and 3.06 mm wide, heavily sclerotized, brownish yellow, glabrous, granulose, each elytron with three and half black spots; 1st spot transverse band like with median elevation and not touching to inner and outer border of elytron; median spot also band like, irregular not touching to outer border and fuses on suture with its counterpart; apical spot small, band like and fuse with counterpart and forms subrounded spot. *Hind Leg (Fig. 60)*; black; coxa 1.05 mm long and 0.50 mm wide, stout; trochanter subtrapezoidal, 0.40 mm long and 0.46 mm wide with straight inner margin, apical corner sharply denticulate; femur broad, 1.94 mm long and 0.71 mm wide, more broader in middle and narrower towards at the apex, outer margin with few fine setae; tibia elongated, 1.77 mm long and 0.31 mm wide, narrow at base and broader at apex, covered by fine dense setae, setae more towards apex; tarsi 4 segmented, 1.43 mm long; 1st segment 0.71 mm long, narrow at base and broader at apex, covered by setae on outer margin; 2nd 0.54 mm long, triangular, dilated and covered by setae on outer margin; 3rd segment very small, quadrate, 0.10 mm long; terminal segment 0.61 mm long, elongated gradually broader towards apex and terminate into bifid claw; claw 0.16 mm long; curved, pointed and with basal tooth.

Abdomen

4.05 mm long and 4.55 mm wide, dorsum black, 8 terga visible, smooth and without hairs; ventral side hairy, black, 6 sternite visible; lateral border covered by setae; coxal line complete; femoral line angulate and terminate very close to antero-lateral side of 1st abdominal sternite, area around the femoral line brownish black and sparsely punctured. *Female genitalia* ; 9th pleurite and a pair of hemisternites together form ovipositor; hemisternites somewhat triangular with pair of stylus. *Receptaculum seminis* ; strongly "U"

PLATE - 8

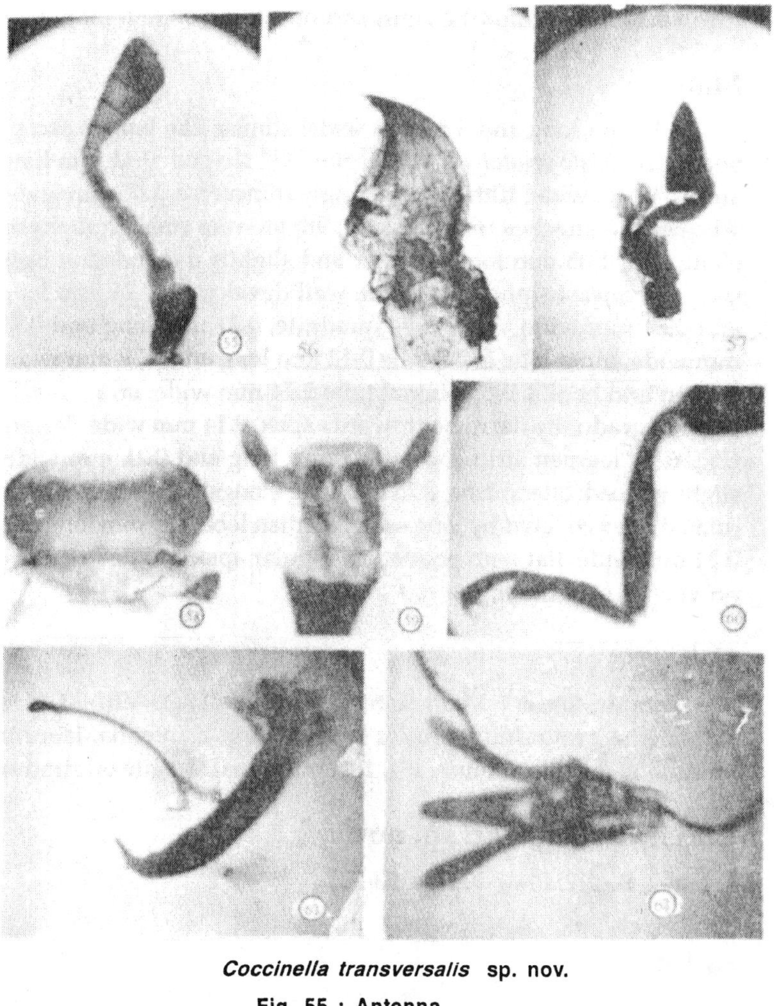

Coccinella transversalis sp. nov.

Fig. 55 : Antenna
Fig. 56 : Mandible
Fig. 57 : Maxilla
Fig. 58 : Labrum
Fig. 59 : Labium
Fig. 60 : Hind leg
Fig. 61 : Last terga of male
Fig. 62 : Tegmen

shaped, broad at base and narrow at apex, marking on surface, cornu 0.20 mm, ramus 0.20 mm and nodulus 0.07 mm long.

Male

5.30 mm long and 3.84 mm wide, similar like female except body size. *Male genitalia :* 9th tergite "U" shaped, 0.41 mm long and 1.79 mm wide; 10th tergite stongly transverse, 1.05 mm wide; 9th pleurite attaches to 9th tergite; 9th sternite small; apophysis elongated, 1.05 mm long, slender and slightly expanded at base (*Fig. 61*). *Sipho :* siphonal capsule well developed, 0.21 mm long and 0.81 mm wide, outer lobe quadrate, 0.21 mm long and 0.51 mm wide, inner lobe of capsule 0.12 mm long and 0.29 mm wide, curved and hook like; siphonal tube 2.74 mm wide, immediately curved, gradually narrower towards apex, 0.14 mm wide. *Tegmen (Fig. 62) :* tegmen strut short, 0.85 mm long and 0.21 mm wide, slightly broad; lateral lobe, 0.75 mm long elongated, narrowed with dilated apex covered by long setae; median lobe 0.75 mm long and 0.21 mm wide, flat with pointed triangular apex and deep groove on ventral side for sipho.

Holotype

Female, India : Maharashtra, Miraj coll., 20-VIII-93, Y.A. Bhosale on groundnut (*Arachis hypoqae* Linn.), antenna, labrum, mandible, maxilla, labium, leg, 1st abdominal sternite on strades.

Coccienlla madhuri **sp. nov.**

(Plate-2, Fig.12; Plate-9, Figs. 63-71)

Female

Body shortly elongate oval, 6.1 mm long and 4.76 mm wide; drosum strongly convex; pronotum 1.70 mm long black; elytra 5.76 mm glabrous with three and half black spots; spermatheca 0.34 mm long with weak markings on surface.

Head

Black, 1.64 mm long and 1.84 mm wide, somewhat squarish, dorso posterior side covered by hair; frons black, somewhat flat, frontoclypeal suture not distinct; clypeus projected anteriorly and expanded infront of compound eyes; surface of eyes smooth with rounded structure, inter ocular distance 1.05 mm; gular sutures not distinct, runs parallel with each other; tentorium with parallel arms. *Antenna (Fig. 63)* : 11 segmented, 1.32 mm long, clavate, brown with black tinge on scape and terminal segment; scape large, broad and some what squarish with 7 long unequal fine setae; pedicel somewhat large, squarish with 4 fine setae on anterior side; 1st segment longer than width; 1st to 6th segments of flagellum, each shows 2 long fine setae; 7th to 9th segments large, 7th with single seta, 8th with 2 fine setae, 9th segment with 5 long setae and apical portion covered by dense fine hairs.

Flagellar formula : SL/W = 2.3, PL/W = 1.75, SL/W = 1.25, 7L/W = 8.3, 8L/W = 0.75, 9L/W = 0.77.

Labrum : flat, 0.44 mm long and 0.78 mm wide, somewhat quadrate with rounded lateral sides, posterior side notched. *Mandible (Fig. 66)* : subtriangular with broad base, 0.58 mm long and 0.50 mm wide at base; apical tooth bifid with depression 0.18 mm; basal tooth somewhat curved and pointed membranous prostheca 0.27 mm. *Maxilla (Fig. 64)* : cardo 0.25 mm long and 0.29 mm wide, as broad as stipe; stipe 0.46 mm long and 0.29 mm wide, gelea and lacinia 0.18 mm long and covered by dense setae on inner side, lacinia flat, leaf like; maxillary palpus 4 jointed, 1.30 mm long, basal joint small 0.08 mm long, 2nd longer than 3rd segment, 0.35 mm long, 3rd segment 0.23 mm long, terminal segment large, 0.75 mm long and 0.73 mm wide, sides divergent, apically truncate, securiform. *Labium (Fig. 65)* : submentum fused with gula; mentum broad and flat, 0.21 mm long and 0.25 mm wide; prementum triangular, 0.27 mm long and wide, outer. border straight with anterior rounded corners and densly covered by fine hairs; labial palpus 3 segmented, 0.44 mm long, basal segment small, 0.06 mm; 2nd segment 0.16 mm long and 0.10 mm wide, more broader than apical segment; apical segment 0.21 mm long and 0.08 mm wide, rounded and bluntly tapering.

Thorax

1.51 mm long and 3.84 mm wide; prothorax broader than long, convergent and inclined anteriorly, anterior margin broadly excavated and cover the head; pronotum black with yellow orange spots on anterior corners, 1.70 mm long and 3.05 mm wide, glabrous; hypomeronflat; middle coxal cavity opened outwardly; hind coxal cavity oval and strongly transverse; dorsum of metathorax concealed by elytra; metendosternite "Y" shaped with broad stalk and paired lateral arms of fulcra; scutellum black. *Elytra (Fig. 12)* : 5.76 mm long and sparsely punctured; elytral suture black; each elytron with three and half black spots, anterior spot small, sub rounded and slightly towards outer margin, median spot irregular band like and fuse on suture with its counterpart, apical spot small, completely cover the apical portion and fuse with its counterpart on suture, near scutellar region half spot which fuse on suture and forms sub rounded spot. *Hind leg (Fig. 67)* : coxa 1.32 mm long and 0.46 mm wide, slender; trochanter 0.46 mm long and 0.48 mm wide, trapeziodal with denticulate process on inner side; femur 1.89 mm long and 0.69 mm wide, broad in middle with fine setae on outer surface; tibia elongated, 1.75 mm long and 0.27 mm wide, cylindrical and covered by dense setae on inner side; tarsi 3 segmented, 1.47 mm long, 1st segment elongated, 0.71 mm long, broad at distal and woth dense coating of setae on outer border; 2nd segment 0.59 mm long slightly rounded to distal end and covered by dense setae on outer side; terminal segment elongated, 0.73 mm long, broad at apex and narrow at base with bifid curved claw and few fine setae; claw 0.14 mm long with basal tooth.

Abdomen

4.1 mm long and 4.53 mm wide, dorsum black, glabrous, 8 terga visible; ventrum black, pubescent, 6 sternites visible; lateral borders covered by setae; coxal line on 1st abdominal sternite complete; femoral line angulate and not reaching to antero lateral side of 1st sternite, area around femoral line hairy. *Female genitalia* : 9th tergite, pair of 9th pleurite and a pair of hemisternite form ovipositor; hemisternite elongated and triangular with stylus. *Receptaculum seminis (Fig. 71)* : somewhat "U" shaped, weak markings on surface, cornu 0.10 mm, ramus 0.24 mm and nodulus 0.08 mm.

Male

5.54 mm long and 4.72 mm wide, same as female except body size. *Male gentalia :* 9th tergite 0.21 mm long and 2.00 mm wide, "U" shaped; 10th terga strongly transverse, 0.29 mm long and 0.94 mm wide, densely covered by long setae; 9th pleurite attached to 9th tergite; 9th sternite sclerotized; apophysis elongated, 1.18 mm long with narrow apex. *Sipho (Fig. 70) :* siphonal capsule well developed, 0.18 mm long and 0.84 mm wide; outer lobe exactly quadrate, 0.21 mm long and 0.42 mm wide; inner lobe strongly curved, 0.18 mm long and 0.31 mm wide and hook like; siphonal tube 2.85 mm long and 0.10 mm wide, elongated with dilated part at base, 0.16 mm; notch between lobes of siphonal capsule prominent. *Tegmen (Figs. 68, 69) :* tegmen strut 1.05 mm long, articulate to basal lobe, elongated with uniform width except apex slightly dilated, 0.21 mm; lateral lobes elongated, cylindrical, 0.63 mm long parallel to median lobe, apex covered by dense long setae; median lobe slightly longer than lateral lobe, 0.73 mm long with groove on ventral side for sipho, apex of lobe triangular and bluntly pointed; basal piece quadrate, 0.56 mm wide with lateral sides running parallel upwardly to tegmen strut.

Holotype

Female, India : Maharashtra, Miraj, Coll., 22-VIII-93, Y.A. Bhosale, on groundnut (*Arachis hypogae* Linn.), antenna, labrum, mandible, maxilla, labium, 1st abdominal sternite, sipho, tegmen, last sterga of male and spermatheca mounted on slides labelled as above.

Paratype

Female 3, male 2, sex ratio (Male : Female), 1:1.5, data same as holotype, Coll. 22-VIII-93, V.A. Bhosale, India, Maharashtra, Miraj.

Prey

Aphid, *Aphis craccivora* Koch.

Prey Plant

Groundnut, *A. hypogae.*

PLATE - 9

Coccinella madhuri sp. nov.

Fig. 63 : Antenna
Fig. 64 : Maxilla
Fig. 65 : Labium
Fig. 66 : Mandible
Fig. 67 : Hind leg
Fig. 68 : Tegmen (enlarged)
Fig. 69 : Tegmen
Fig. 70 : Sipho
Fig. 71 : Spermateca

Coccinella bhuimungi sp. nov.

(Plate-2, Fig. 14; Plate-10, Figs. 72-79)

Female

Body large, 6.04 mm long and 5.85 mm wide, shortly oval, and strongly convex on dorsal side; pronotus 1.3 mm long and 3.16 mm wide; antenna 1.41 mm long; elytra brownish red, 5.58 mm long and 3.33 mm wide, glabrous, each with two and half spots; receptaculum seminis 0.24 mm long with weak markings on surface.

DISCUSSION

This species runs close to *C. tibetena*, *C. lama*, *C. nigrovittata* and *C. tranversalis* by having antenna, maxilla, labium, pronotum black with brownish yellow spot on anterior margins, scutellum black. It also resembles with *C. transversalis* by having apophysis bifid at base, twisted apex of sipho. But it differs from the above species by having the following characters :

i. Elytra is with two and half spots, elevated margin to the outer border,

ii. Posterior border of labrum straight, lateral border angulate in middle,

iii. Mouth parts yellowish brown,

iv. Area around femoral line on 1st abdominal sternite closely punctured,

v. Depression in the siphonal capsule less, inner lobe slightly curved and bluntly pointed, siphonal tube unifrm in width,

vi. Median lobe of tegment more narrowly bluntly pointed at its apex.

vii. Receptaculum seminis with weak markings on entire surface and

viii. Flagellar formula.

Head

Black, 1.89 mm long and 2.21 mm wide; frons smooth, frontoclypeal suture not distinct; clypeus projected anteriorly and expanded infront of compound eyes; eyes black, rough and shiny, interocular distance 1.35 mm; gular sutures parallel; tentorium with parallel arms; mouth parts yellowish brown with black tinge. *Antenna (Fig. 72)* : 11 segmented, 1.41 mm long, clavate, yellowish brown with black tinge on scape on last three segments; scape large, elongated, squarish with 8 fine setae placed on inner lateral sides, longest seta 0.16 mm; pedicel smaller than scape, longer than other segments with three setae on inner lateral sides; 1st segment longer and wide; segments onward from fourth bend inner side; 7th, 8th and 9th segments large; 7th segment with single seta on inner lateral side; 8th segment with 2 setae; one on each lateral side; 9th segment apically turncate with 9 fine long setae with dense coating of hairs.

Flagellar formula : SL/W-1.68, Pl/W=1.4, 1L/W-2, 5L/W=1.25, 7L/W=0.84, 8L/W=0.63, 9L/W=1.

Labrum (Fig. 75) : somewhat squarish, 0.42 mm long and 0.97 mm wide; lateral borders rounded with angulate portion in the middle and covered by setae; anterior side slightly depressed posterior side straight. *Mandible (Fig. 73)* : quadrate, 0.69 mm long and 0.63 mm wide at base; apical tooth bifid with 0.25 mm depression; basal tooth curved and bluntly pointed; membranous prostheca 0.33 mm wide; stipe 0.46 mm long and 0.25 mm wide; galea and lacinia 0.21 mm long with dense setae on inner side; maxillary palpus four segmented, 1.30 mm long; 1st segment small 0.08 mm; 2nd segment 0.33 mm, narrow at base and broad at apex, longer than 3rd; 3rd segment 0.27 mm; terminal segment 0.75 mm, large, apically truncate and securiform. *Labium (Fig. 76)* ; submentum fused with gula; mentum triangular, 0.21 mm long and 0.31 mm wide; prementum quadrate, 0.27 mm long and 0.42 mm wide, apically bordens by hairs; labial palpus 3 segmented 0.42 mm long, basal segment small, 0.04 mm; 2nd segment 0.16 mm; 3rd 0.21 mm and bluntly pointed.

Thorax

1.68 mm long and 4.00 mm wide; prothorax broader than long; pronotum black, 1.83 mm long and 3.16 mm wide, glabrous, anterior corners angulated with brownish spot; surface of hypomeron flat, slightly foveolate; mesosternum short, metastermum large, hind coxal cavity oval and strongly transverse; metendosternite "Y" shaped. *Elytra (Fig. 14)* : 5.58 mm long and 3.33 mm wide, reddish brown, glabrous, granulate with elevated margin towards outer border; each elytron with two and half black spots, anterior side without spot; posterior spot irregular, band like, not touching to outer border and fuse with its counter part on suture; apical spot band like with small spherical outer part, middle narrow and band like, inner part which fuse with its couter part on suture; half spot on anterior side near pronotum fuses with its counter part on suture and forms small rounded spot on suture; sutre entirely black. *Hind leg (Fig. 77)* ; black; coxa 0.84 mm long and 0.48 mm wide; trochanter 0.42 mm long and 0.52 mm wide, somewhat triangular with denticulate process on inner side; femur broad, 1.98 mm long and 0.75 mm wide, narrow towards apex; tibia 1.85 mm long and 0.27 mm wide cylindrical, basal part narrow, inner side convered by dense, thick, prominant spines; tarsi cryptotetramerous, 2.02 mm long; basal segment 0.73 mm long and 0.25 mm wide, narrow at base and broad at apex with setae on outer side; 2nd segment 0.54 mm long and 0.12 mm wide, rounded apex with setae; 3rd segment very small, 0.10 mm long and 0.08 mm wide; terminal segment 0.65 mm long and 0.16 mm wide, gradually broad towards apex and terminated into bifid claw; claw stout, 0.27 mm curved and pointed with basal tooth.

Abdomen

4.22 mm long and 4.64 wide, dorsal side black, 1-7 terga slightly sclerotized; six sternite visible; ventral side blackish brown, sparsely punctured; coxal line complete; femoral line angulate, area around femoral line closely punctured. *Female genitalia* : 9th tergite, pair of ovipositor, hemisternites with stylus, elongated and sub triangular. *Receptaculum seminis* ; weakly "U" shaped, slightly broad at base, weak markings on surface, cornu and ramus 0.18 mm, similar and nodulus 0.06 mm.

Male

5.35 mm long and 4.04 mm wide, similar like female except body size. *Male genitalia* : 9th tergite "U" shaped, 0.31 mm long and 1.68 mm wide, relatively large; 10th tergite 0.21 mm long and 1.26 mm wide, relatively large; 10th tergite 0.21 mm long and 1.26 mm wide, strongly transverse; 9th pleurite attached to 9th tergite; 9th sternite small; apophysis elongated 1.26 mm long and bifid at base. *Sipho (Fig. 79)* ; 2.84 mm long; siphonal capsule 0.21 mm long and 0.92 mm wide, well developed; outer lobe 0.25 mm long and 0.53 mm wide, quadrate; inner lobe 0.18 mm long and 0.31 mm wide slightly curved and bluntly pointed, depression between lobes not prominent; siphonal tube, 2.63 mm long and 0.10 mm wide, uniform in width, immediately curved with narrow and twisted apex. *Tegmen (Fig. 78)* : tegmen strut elongated, 0.95 mm long, broad at apex and narrow at base; lateral lobes longer than median lobe, 1.16 mm, apex densely covered by setae; median lobe 0.99 mm long, triangular and more narrowed elongated and bluntly pointed at apex.

Holotype

Female, India : Maharashtra, Miraj, Coll., 29-VIII-93, Y.A. Bhosale, on groundnut (*Arachis hypogae* Linn.), antenna, labrum, maxilla, mandible, labium leg, 1st abdominal sternite, spematheca, sipho, tegmen and last abdominal terga on slides labelled as above.

Paratype

Female 1, male 8, sexratio (*Male:Female*), 1:0.125 data same as in holotype and coll. from 20-VIII-93 to 30-VIII-93, Y.A. Bhosale, on groundnut (*A. hypogae*) and subabul (*Luceana retusa* Brew.), India : Maharashtra, Miraj.

Prey

Aphid, *Aphis craccivora* koch.; Psyllid, *Heteropsylla cubana* Crowford.

Prey Plant

Ground nut, *A. hypogaea*, Subabul, *L. retusa*.

PLATE - 10

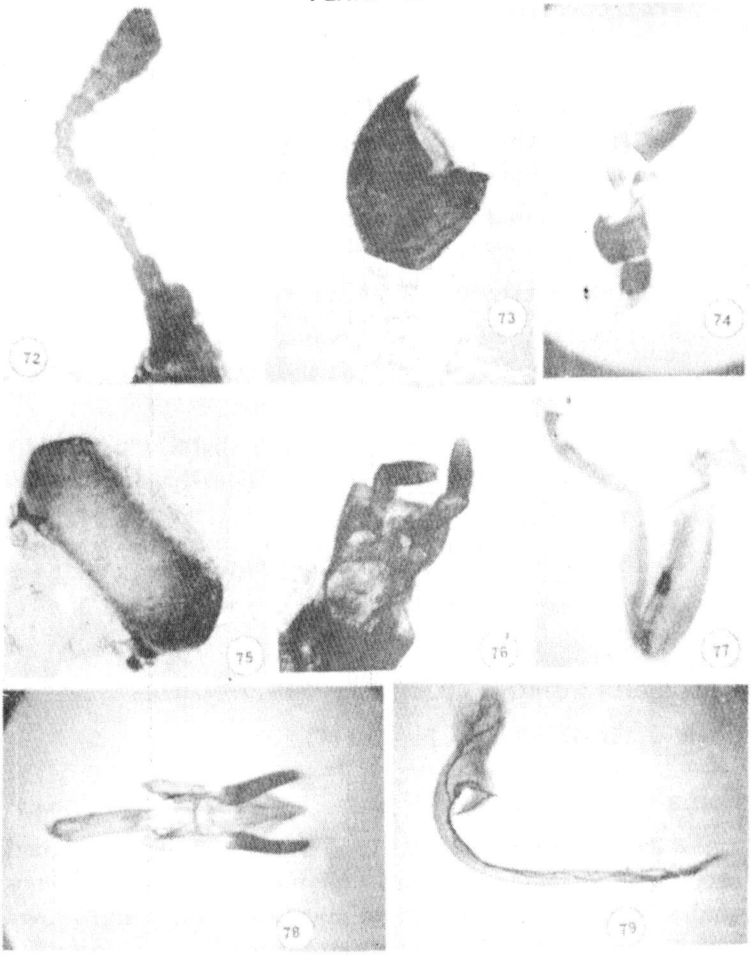

Coccinella bhuimungi sp. nov.

Fig. 72 : Antenna

Fig. 73 : Mandible

Fig. 74 : Maxilla

Fig. 75 : Labrum

Fig. 76 : Labium

Fig. 77 : Hind leg

Fig. 78 : Tegmen

Fig. 79 : Sipho

DISCUSSION

The species resembles to *C. tibetina*, *C. lama*, *C. nigrovittata* and *C. transversalis* by having head black, body sub-ovate, pronotum black, antenna 11 segmented, clavate. It also resembles with *C. tibetina* by having last tergite of male. It also resembles with *C. transversalis* by having same coxal line and femoral line. However, differs from the above species having the following characters :

(i) Elytra glaborus and punctate, each elytra with three and half black spots, anterior spot small subrounded,

(ii) Antennal scape with 7 long unequal setae,

(iii) Notch between lobes of siphonal capsule very prominent, inner lobe strongly curved and hook like, apex of sipho not twisted,

(iv) Lateral lobes of tegmen brownish, slightly curved to the apex side, median lobe slightly longer than lateral lobes.

(v) Apophysis not bilobed at base.

(vi) Area around femoral line hairy and

(vii) Flagellar formulla.

Coccinella yaminii sp. nov.

(Plate -2, Fig. 15; Plate-11, Figs. 80-87)

Female

6.29 mm long and 5.73 mm wide, body broadly oval; pronotum 1.73 mm long, glabrous; elytra 5.0 mm long castaceous; ventral side black; antenna 1.34 mm long, clavate and brown; supermatheca 0.29 mm long with strong markings on surface.

Head

Black, 1.79 mm long and 2.04 mm wide, somewhat quadrate with rounded lateral borders; frons smooth; clypeus expanded, fronto clypeal suture indistinct; interocular distance 1.47 mm; gular sutures indistinct; mouth parts brown with black tinge. *Antenna (Fig. 80)* : 11 segmented, 1.34 mm long clavate; terminal segment with black tinge; segments onwards from 6th bend inwardly; scape large, longer than width, without setae; pedicel shorter than scape, somewhat squarish and without setae; 1st segment longer than width; 2nd to 6th segments of flagellum more or less equal, without setae; last 3 segments large, 8th segment with only one seta on inner antero-lateral side.

Flagellar formula : SL/W=1.16, PL/W=1, 1L/W=2, 5L/W=1.25, 7L/W=0.75, 8L/W=0.55, 9L/W=0.76.

Labrum : completely sclerotized, 0.42 mm long 0.93 mm wide, lateral sides rounded with slight elevations in the middle, side towards clypeus straight, anterior side with long setae, setae more on antero-lateral side. *Mandible (Fig. 81)* : somewhat quadrate, 0.67 mm long and 0.52 mm wide; apical tooth bifid, depression between bifid tooth 0.20 mm; basal tooth broad at base and bluntly pointed; mandible base broad; membranous prostheca 0.23 mm *Maxilla (Fig. 83)* : cardo somewhat rounded, 0.27 mm long and 0.31 mm wide, as wide as stipe; stipe 0.51 mm long and 0.31 mm wide; galea 0.31 mm long and lacinia 0.31 mm long, elongated with dense coating of setae on inner side; maxillary palpus 4 segmented, 1.37 mm long; basal segment very short, 0.10 mm long; 2nd segment 0.40 mm long with single seta on inner side; 3rd segment 0.29 mm long; terminal segment large, 0.75 mm long and 0.73 mm wide, apically truncate, securiform. *Labium (Fig. 82)* : submentum fused with gula; mentum trapezoidal, 0.21 mm long 0.31 mm wide; prementum subquadrate, 0.29 mm long and 0.31 mm wide; anteriorly covered by numerous hairs; labial palpus 3 segmented, 0.44 mm long; basal segment very short, 0.10 mm; 2nd and 3rd segments more of less elongated, 0.16 mm long, 3rd segment 0.18 mm long.

Thorax

1.79 mm long and 4.11 mm wide; prothorax convergent and inclined anteriorly; pronotum black, 1.73 mm and 3.16 mm wide, glabrous with two yellow spots on the antero-lateral side, antero-middle side weakly convex, antero-lateral corners sub angulate and postero-lateral corners are rounded; hypomeron broad, posterior-part strongly extends behind coxa; front coxal cavity distinclty separated; mesosternum short with median psoterior inter coxal process; metasternum nearly rectangular with interior intercoxal process; articulation of meso- and metasternum compact;

middle coxal cavity narrowly opened outwardly;hind coxal cavity transversely oval; metendosternite with paired lateral arms of fulcra; scutellum black. *Elytra (Fig. 15)* : 5.00 mm long and 3.18 mm wide, castaceous, heavily sclerotized glabrous and sparsely punctured; epipleuron horizontal and smooth; each elytron with three black spots, anterior spot irregular, band like, not touching to outer margin and elevated in the middle; median spot thick, band like, zigzag, not touching to outer margin but fuse on suture with its counter part on another elytron; apical spot small, touching to outer border and fuse on suture with its counter part; suture black, sub rounded black spot on the scutellar region. *Hind leg (Fig. 85)* : black, coxa 1.05 mm long and 0.52 mm wide, elongated and broad; trochanter somewhat triangular, 0.42 mm long and 0.48 mm wide with denticulate process; femur elongated, 1.89 mm long and 0.78 mm wide, broad through out, setae on inner side, setae more concentrated towards apex; tibia elongated, 1.89 mm long and 0.44 mm wide, cylindrical, narrow at base, gradually broad towards apex, thick setae on inner side; tarsi 4 segmented, 1.47 mm long, 1st segment elongated, 0.73 mm long gradually broad towards apex with uniform coating of setae on inner side; 2nd segment 0.56 mm long narrow at base with setae on inner and anterior side; 3rd segment small. 0.08 mm long; terminal segment elongated, 0.69 mm long with broad apex; claw bifid, 0.25 mm long, gradually curved and slightly pointed with basal tooth.

Abdomen

4.11 mm long and 4.64 mm wide, terga of 1- segment weakly sclerotized; 8th tergite strongly sclerotized; six sternite visible; femoral line incomplete and angulate, area around femoral line blackish brown and sparsely punctured (Fig. 84); coxal line complete; ventrum blackish, sparsely punctured. *Female genitalia* : hemisternite with stylus; stylus elongated, triangular. *Receptaculum seminis* : strongly "U" shaped braod at base, narrow at apex, strong markings on surface; cornu 0.21 mm long and nodulus 0.08 mm.

Male

0.97 mm long and 4.04 mm wide smaller than female. *Male genitalia* : 9th tergite "U" shaped, 0.35 mm long and wide, large;

10th tergite strongly transverse, 0.10 mm long and 1.56 mm wide; 9th pleurite fuse to 9th tergite; 9th sternite small; appophysis elongated, 1.16 mm long, base bifid. *Sipho (Fig. 87) :* siphonal capsule well developed, 0.21 mm long and 0.84 mm wide, outer lobe quadrate, 0.21 mm long and 0.46 mm wide inner lobe curved blunt 0.24 mm long and 0.35 mm wide and blunt; siphonal tube immediately curved, 2.63 mm long and 0.18 mm wide, wide twisted. *Tegmen (Fig. 86) :* tegmen strut 0.82 mm long and 0.21 mm wide, apex broad elongated; lateral lobes 0.65 mm long and 0.16 mm wide, parallel to median labe; median lobe 0.67 mm long and 0.35 mm wide, slightly longer than lateral lobes, gradually tapered and bluntly pointed.

Holotype

Female, India : Maharashtra, Miraj, Coll., 23-VIII-93, Y. A. Bhosale, on groundnut (*Arachis hypogae* Linn.), antenna, labrum, mandible, maxilla, labium, 1st abdominal sternite, spermatheca, sipho, tegmen and last abdominal terga of male on slides labelled as above.

Paratype

Female 40, male 26, sex ratio (males:female), 1:1.53 data same as in holotype coll., from 23=III-93 to 31-VIII-93, Y.A. Bhosale, India : Maharashtra, Miraj.

Prey

Aphid, *Aphis craccivora* Koch.

Prey Plant

Groundnut, *A. hypoqae.*

DISCUSSION

This species close to *C. tibetena, C. Lama, C. nigrovittata* and *C. transversalis* by having similar maxilla, labium, head black, body ovate, pronotum black with yellowish brown spot on anterior margins, scutellum black, antenna 11 segmented clavate. It resembles with *C. tibetena* by having median lobe of tegmen in male

PLATE - 11

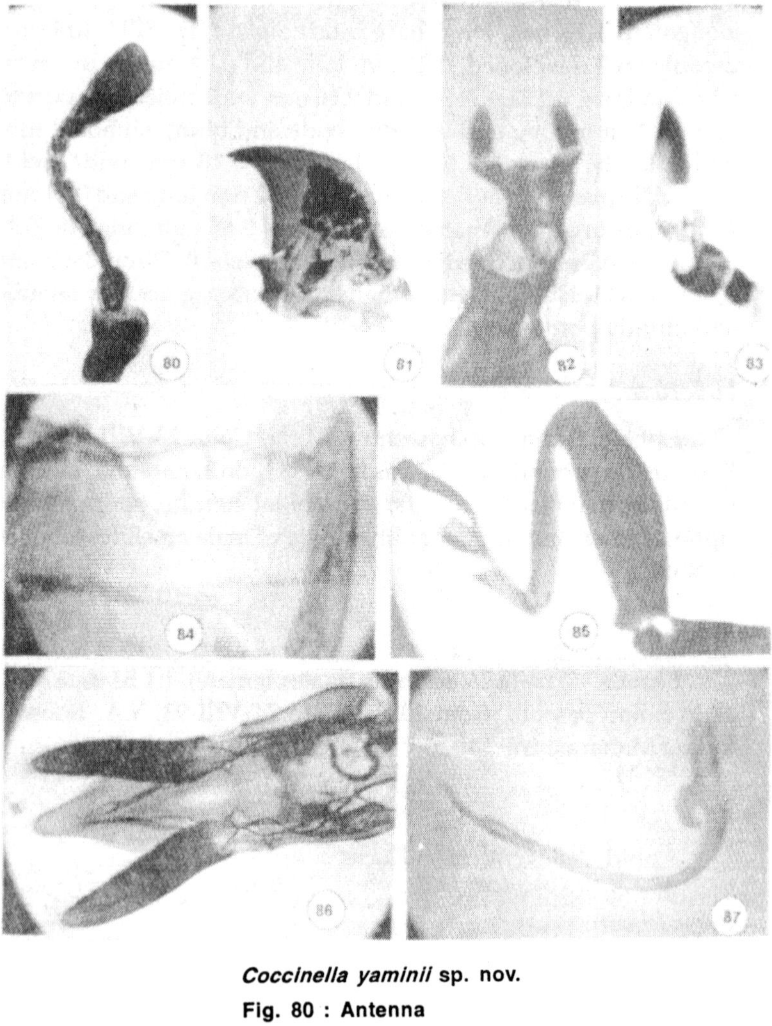

Coccinella yaminii sp. nov.

Fig. 80 : Antenna

Fig. 81 : Mandible

Fig. 82 : Labium

Fig. 83 : Maxilla

Fig. 84 : 1st abdominal sternite

Fig. 85 : Hind leg

Fig. 86 : Tegmen

Fig. 87 : Sipho

gradually narrowed with rounded apex and with C. *transversalis* by having coxal line complete, femoral line angulate and same elytral spots. However, it differs from the above species by having the following characters :

i) Elytra castaceous, glabrous, sparasely punctured and granulate,

ii) Mouth parts yellowish brown with black tinge,

iii) Antenna without setae except last segments and bend from 6th segment onwards,

iv) Spermatheca exactly "U" shaped with broad base, with strong markings on surface.

v) Inner lobe of siphonal capsule slightly curved and not pointed but blunt, siphonal tube immediately curved and gradually narrowed with twisted apex,

vi) Base of apophysis not bifid,

vii) Area around femoral line black and sparsely punctured and

viii) Flagellar formula.

Coccinella yogini sp. nov.

(Plate-2, Fig. 29; Plate-24, Figs. 183-186)

Female

13 mm long and wide, body large, strongly convex on dorsal side; pronotum 4.22 mm long and 7.59 mm wide, glabrous; elytra 5.77 mm long and 6.50 mm wide, reddish brown with six full and two half spots; antenna 2.15 mm long, yellowish; spermatheca 2.01 mm long with markings on surface.

Head

Reddish, elongately oval, 3.01 mm long and 3.27 mm wide; frons reddish, fronto clypeal suture indistinct; clypeus projected anterioly; eyes connected to posterior side by black patches, interocular distance 1.68 mm; posterior border black; ventral side with distinct prallel gular sutures. *Antenna (Figs. 183, 184)* : 11

segmented, 2.15 mm long, serrated, yellowish with black tinge on the terminal segment; scape large, broad at apex, rounded lateral borders with 4 to 6 long seate; pedicel quadrate with elevated portion, outer lateral border with 3 long setae; 1st to 6th segments more or less equal in length and each with 3 to long setae; 7th and 8th large, somewhat similar with prominent serrated structure and covered by setae; 9th segment large, rounded with black tinge and covered by setae; 3rd segment onwards the serrated structure prominent.

Flagellar formula : SL/W=1.2, PL/W=1.3, 1L/W=1.5, 5L/W=1.1, 7L/W=1.0, 8L/W=0.72, 9L/W=1.1.

Labrum (Fig. 186) : yellowish, 0.61 mm long and 1.30 mm wide, squarish, antero-lateral borders rounded and densely covered by long thick setae, posterior border flat. *Mandible*, large, 1.05 mm long and 0.61 mm wide, quadrate; apical tooth bifid with 0.35 mm depression; basal tooth sharp and small; membranous prostheca 0.52 mm; base of mandible 0.92 mm wide and highly sclerotized. *Maxilla (Fig. 185)* : cardo 0.85 mm long and 0.83 mm wide, rounded, slightly broader than stipe; stipe elongated, 0.98 mm long and 0.73 mm wide, spatulate and densely covered by setae 0.62 mm long; lacinia simple. 0.62 mm long, flat with dense thick setae; maxillary palpus 4 segmented; 1st segment small, 0.20 mm long; 2nd and 3rd segments more or less equal, 0.08 mm long; terminal segment large; 0.80 mm long, truncate apically and securiform. *Labium* : gula and submentum fused; mentum quadrate, 0.63 mm long and 0.45 mm wide; prementum rounded, 0.35 mm long with setae on apical side; labial palpus 3 segmented; 1st very small. 0.18 mm; 2nd large, 0.35 mm and terminal segment elongated, 0.53 mm long and roundly ended.

Thorax

2.00 mm long and 6.89 mm wide; pronotum reddish brown, glabrous with 2 large black spots on the centre, black spots fuse with each other at base and slightly separated at apex, antero-lateral borders arcuate, postero-lateral borders rounded, anterior border slightly elevated in middle, dorsum convex in middle; metasternum rectangular, middle coxal cavity outwardly open; hind coxal cavity oval; scutellum small, black and triangular. *Elytra*

(Fig. 29) : reddish brown, 5.77 mm long and 6.50 mm wide, glabrous sparsely punctate at base, larger than pronotum, antero-lateral border more angulate, inner border strongly inclined, middle portion strongly convex; outer border comes out strongly and flat, looks like tortoise shell; each elytron with six full and two half spots; in the middle 4 spots, basal spot large and subrounded, middle 2 spots subrounded and placed on the equal distance, apical spot small and covers the tip of elytron; outer portion of elytrons with 2 subquadrate black spots, these placed one between first and second middle spot, and second between second and third middle spots, both these spots touching to outer border of elytron; inner region with two half spots touching to inner border, these inner spots fuse on suture with its counter part on another elytron, inner upper spot somewhat triangular large with tip portion towards scutellum, another inner spot small and subrounded; elytral epipleuron reaching up to apex, more flat in middle and without fovae. *Hind leg* : coxa 1.83 mm long and 0.93 mm wide, strong and stout; trochanter tirangular, 0.65 mm long; femur elongated, 2.95 mm long and 1.35 mm wide; tibia 2.85 mm long and 1.25 mm wide long and densely covered by thick tibial spines; tarsi 3 segmented, 3.00 mm long elongated; 1st and 2nd segments distaly rounded and covered by setae; terminal segment narrow at base, gradually broad towards apex and terminate into claw; claw 1.09 mm long, bifid, pointed with basal tooth.

Abdomen

Reddish, 7.8 mm long and 8.00 mm wide; six sternites visible; coxal line complete; femoral line incomplete and slightly angulate, area around femoral line brownish an closely punctured. *Female genitalia* hemisternites with elongated stylus. Receptaculum seminis "U" shaped, 2.01 mm long, markings on surface.

Male

Unknown.

Holotype

Female, India : Maharashtra, Kolhapur, coll. 4-IV-94, T.V. Sathe, on Cotton (*Gossypium herbaceum* Linn.), antenna, labrum, mandible, maxilla, labiu, hind leg on slides labelled above.

Paratype

Female 1, data same as in holotype coll., 4-IV-94, India : Maharashtra, Kolhapur.

Prey

Aphid, *Aphis gospii* Koch.

Prey Plant

Cotton *G. herbaceum.*

DISCUSSION

This species resembles to *C. transversalis* by having similar mandible, maxilla, labium, scutellum black, coxal and femoral line on 1st abdominal sternite. However, it differs from the above species by having following characters :

i) Body large, tortoise shell like elytra,

ii) Elytral spots, each elytron with six full and two half spots.

iii) Antenna serrated, scape and pedicel with large setae,

iv) Pronotum with 2 black spots on middle and

v) Flagellar formula.

Genus *Coelophora* Mulsant, 1850

The genus *Coelophora* is erected by Mulsant in 1850. Under this genus 9 species have been described. *Coelophora inaequlis* Fabricius, 1775; *C. unicolor* Fabricius, 1792; *C. reniplagiata* Mulsant, 1850; *C. bissellata* Mulsant, 1850; *C. sexereata* Mulsant, 1853; *C. faivrei* Mulsant, 1866; *C. formosa* Crotch, 1874; *C. deficiens* Crotch, 1874 and *C. nitidicollis* Kapur, 1924; *C. inaequalis* var. *novem-punctata* Fabricius, 1775; *C. inaequalis* var. *iridea* Thunberg, 1781; *C. inaequalis* var *psi* Thunberg, 1781; *C. inaequalis* var. *partita* Mulsant, 1850; *C. inaequalis* var. *mastersi*, Blackburn, 1892; synonemised species includes :

C. unicolor (Fabricius) Mulsant, 1850.

Annls. Soc. Agric. Lyon 2, 413.

C. unicolor (Fabricius) var. *calignousa* Mulsant; Weise 1892.

Annls. Soc. ent. Belg. 36, 21.

C. unicolor (Fab.) var. *cinctipennis* Weise, 1892.

Annls. Soc. ent. Belg. 36, 20.

C. unicolor (Fab.) var. *signata* Weise, 1892.

Annls. Soc. ent. Belg. 36, 21.

C. unicolor (Fab.) : Korschefsky, 1932.

Coleopt, Cat. 16, 297.

C. sexareata Mulsant : Korschefsky, 1932.

Coleopt. Cat. Berl. Pt. 120, Coccinellidae 2 : 29

C. bissellata (Mulsant) : Korschefsky, 1932.

Coleopt. Cat. Berl. pt. 120, Coccinellidae 2 : 291.

The genus is visualized by following characters : body subcircular, glabrous and strongly convex; antenna thin, longer than frons; interocular distance about half the head width; lateral sides and basal corners of pronotum gently rounded; prosternum carinate; scutellum broad, triangular; elytral base much broader than pronotum; first abdominal sternum with incomplete femoral line; middle and hind tibia with two small spurs.

Coelophora orientalis sp. nov.

(Plate-2, Fig. 16; Plate-12, Figs. 88-95)

Male

5.01 mm long and 4.90 mm wide, semicircular, yellowish red; pronotum 0.14 mm long, punctate with 4 black spots; elytra 4.80 mm long yellowish red, glabrous with four full and two half black spots; siphonal tube 4.30 mm long, curved.

Head

Somewhat round, postero-lateral broder black, 1.43 mm long and 1.73 mm wide, inserted into pronotum and not visible from outside; frons pale yellowish, fronto clypeal sutre indistinct; clypeus slightly projected anteriorly and expanded in front of compound eyes, anterior margin straight; compound eyes projected antero-laterally with shiny black suface, interocular distance 0.82 mm; gular sutures prominent and runs parallel, tentorium with parallel arms; mouth parts brown. *Antenna (Fig. 88)* : 11 segmented, 1.54 mm long, clavate with less prominent club, yellowish brown, thin with black tinge on terminal segment; scape large, apically broad with 3 long fine setae; pedicel slightly longer than width with 2 fine setae; 1st five segment longer than width and more or less equal with single seta; 6th segment smaller; 7th to 9th segment forms less compact club; 7th segment shows single long seta and few fine setae; 8th segment transverse with 2 long setae and few fine setae; terminal segment elongated oval with 8 long setae, 2 on each lateral side and 4 on apex with numerous small fine setae on anterior border.

Flagellar formula : SL/W=2.25, PL/W=1.4, IL/W=2, 6L/W=1.25, 7L/W=1.2, 8L/W=0.83, 9L/W=1.5.

Labrum (Fig. 91) : somewhat quadrate, 0.65 mm long and 0.90 mm wide, plate like with rounded lateral sides and without setae. *Mandible (Fig. 90)* : sub triangular, base broad, 0.5 mm long and 0.26 mm wide at base; apical tooth bifid with depression 0.22 mm; basal tooth small, slightly curved and bluntly pointed; membranous prostheca 0.21 mm. *Maxilla (Fig. 89)* ; cardo 0.25 mm long and wide; stipe longer than width, 0.40 mm long and 0.30 mm wide; galea 0.27 mm long, large, hook like with coating of setae; lacinia 0.25 mm long, leaf like with setae; maxillary palpus 4 segmented, 1.24 mm long, basal segment small. 0.16 mm long; 2nd longer than 3rd, 0.31 mm long; 3rd as long as broad, 0.21 mm long with setae, terminal segment large, 0.54 mm long and 0.50 mm wide apically truncate and securiform. *Labium (Fig. 92)* : post mentum fused with gula; mentum large, 0.25 mm long and wide; prementum quadrate, 0.13 mm long and broad with rough surface and apically not borders by hair; labial palpus 3 segmented, 0.35 mm long, basal segment small, 0.06 mm long; 2nd 0.12 mm, apically broad, 3rd segment elongated, 0.16 mm long and roundly ended.

Thorax

1.37 mm long and 2.31 mm wide, pronotum broader than long, 0.14 mm long and 2.95 mm wide, densely punctate in middle, yellowish with black spots along the basal margin, outer pair small touch to border of pronotum, inner pair large, subtriangular place close with each other; antero lateral corners somewhat pointed and lateral and basal corners gently rounded, anterior side of pronotum with depression with slightly concave surface in middle; prothoracic hypomeron with round shallow depression on anterior side; prostermum with pair of carinae; mesosternum deeply and triangularly emarginate in middle; scutellum broad and triangular. *Elytra (Fig. 16)* : 4.80 mm long and 3.26 mm wide, base much broader than pronotum, yellowish brown, glabrous, densely punctate towards outer margin and sparsely punctured towards inner side; each elytron with four full and two half black spots; scutellar spot half, small subtriangular and fuse on suture with its counterpart on another elytron; basal spot small, subrounded and found towards the anterolateral side of elytron; median spot somewhat large, sub quadrate; on lateral side of median spot a small, somewhat squarish spot touching to the outer border of elyton; posterior side of elytron with small, subrounded black spot; apical spot half, fuse on suture with its counterpart on another elytron; elytron; elytron; elytral suture not black; elytral epipleuron entire, strongly descending externally. *Hind leg (Fig. 93)* ; coxa 1.18 mm long and 0.46 mm wide, stout; trochanter 0.58 mm long and 0.13 mm wide, sub quadrate with depression on inner lateral side are with denticulate process with long fine setae; femur large, 1.89 mm long and 0.73 mm wide, broad in middle with setae on inner side; tibia elongated, 1.89 mm long and 0.33 mm wide, cylindrical with prominent curved pair of tibial spurs 0.12 mm long, tibial spines on inner lateral side; trasi 4 segmented, basal segment large, 0.69 mm long, triangular with coating of setae; 2nd elongated, 0.69 mm long with thick covering of setae; 3rd very small, 0.04 mm long, quadrate, 4th elongated, 0.69 mm long, gradually broad towards apex and terminated into bifid claw; claws 0.18 mm, pointed, curved, long with very small basal tooth.

Abdomen

2.53 mm long and 2.84 mm wide, six sternites visible, ventrum yellowish pubescent; coxal line on 1st abdominal sternite complete, femoral line incomplete and slightly angular; lateral borders covered by hairs. *Male genitalia* : 9th tergita "U" shaped; 10th tergite elevated, short, transverse and covered by setae; apophysis straight and without bilobed base. *Sipho (Fig. 94)* : 4.40 mm long; siphonal capusle well developed, 0.10 mm long and 0.63 mm wide, outer and inner lobe inwardly curved, outer lobe 0.12 mm long and 0.13 mm wide; inner lobe 0.06 mm long and 0.31 mm wide, pointed and more curved than outer lobe, base with slight elevated portion, depression between two lobes prominent; siphonal tube 4.30 mm long, immediately curved, apex digitate and terminate pointly. *Tegmen (Fig. 95)* : lateral lobes 0.88 mm long, as long as median lobe, apex dilated, middle slightly curved, apex covered by long setae; median lobe 0.88 mm long, flat broad at base, apex more narrowed and bluntly pointed; tegmen strut 1.37 mm long, broad at apex.

Female

Unknown.

Holotype

Male, India : Maharashtra, Kolhapur, coll., 8-VIII-93, Y.A. Bhosale, on Compositae garden plant (*Gallardia pulchella* Fouger), antenna, labrum, mandible, maxilla, labrum, hind leg, sipho and tegmen on slides labelled as above.

Paratype

Male 2, data same as in holotype, coll., from 8-VIII-93 to 12-VIII-93, Y.A. Bhosale, India : Maharashtra, Kolhapur.

Prey

Aphid, *Dactynotus compositae* Theobald.

Prey Plant

Compositae garden plant, *G. pulchella*.

PLATE - 12

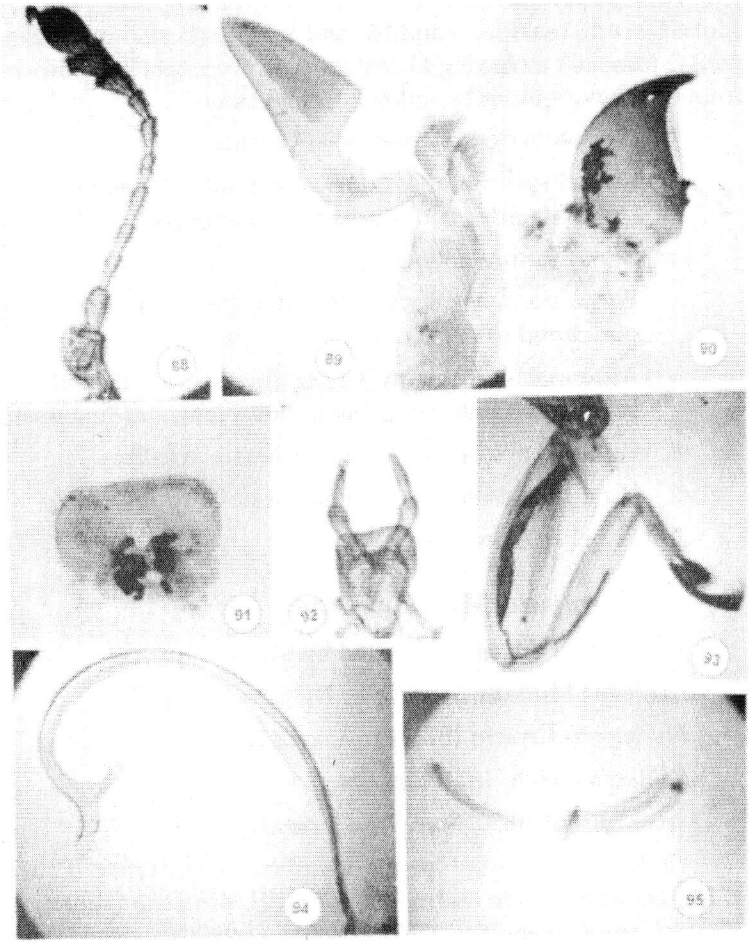

Coelopora orientalis sp. nov.

Fig. 88 : Antenna

Fig. 89 : Maxilla

Fig. 90 : Mandible

Fig. 91 : Labrum

Fig. 92 : Labium

Fig. 93 : Hind leg

Fig. 94 : Sipho

Fig. 95 : Tegmen

DISCUSSION

This species resembles to *C. bissellata* by having same pronotal spots, labrum, maxilla, mandible and tegmen. It also resembles with *C. inaequalis* by having 4 basal spots on pronotum But it differs from the above species by following characters :

i. Pronotum densely punctate in middle,

ii. Elytral spot found on the lateral side of median spot small, squarish and touched to outer border of elytron,

iii. Elytral suture not black,

iv. Elytra densely punctate to outer margin and sparsely punctured to inner side,

v. Antennal scape with 3 long fine setae, 7th and 8th segment on their lateral side shows long and fine setae,

vi. Siphonal apex digitate and terminate pointly.

vii. Hind tibia with thick prominent tibial spurs and

viii. Flagellar formula.

Genus *Harmonia* Mulsant, 1850

The genus *Hermonia* is erected by Mulsant in 1850.

Harmonia Mulsant, 1850, *Spec. Trim. Securipalp*: 74, 75.

Ptychanatis Crotch, 1874, *Rev. Cocci.*, 122.

Callineda crotch, 1871. *Cat. Cocci.*, 6.

Leis Mulsant, 1850, *Spec. Trim Securipalp.*, 241.

Under this genus the species described are *H. axyridis* (Pallas, 1773); *H. octomaculata* (Fabricius, 1781); *H. dimidiata* (Fabricius, 1781); *H. arcuata* (Fabricius, 1787), *H. sedecim-notata* (Fabricius, 1801).

The genus is represented by following characters : body oval, longer than broad, moderately convex and medium sized; antenna longer than frons, 8th segment transverse and strongly broadening apically; lateral sides of pronotum strongly arcuated and marginated while basal corners rounded; prosternal process with or without two carinae; scutellum triangular; elytral base wider than pronotum; first abdominal sternum with an incomplete femoral line; tibial spurs indistinct.

Harmonia soyabinii sp. nov.

(Plate-2, Fig. 17; Plate-13, Figs. 96-104)

Female

Body elongately oval, 6.43 mm long and 4.43 mm wide, weakly convex on dorsal side; pronotum 2.00 mm long and 4.22 mm wide, glabrous and without spot; antenna 1.58 mm long; elytra 4.22 mm long and 2.20 mm wide with four and half black spots; spermatheca "V" shaped, 0.82 mm long and 0.18 mm wide.

Head

1.62 mm long and 2.25 mm wide, somewhat quadrate with rounded lateral sides; frons flat and smooth with 2 small rounded black spots on posterior side; frontoclypeal suture not distinct; clypeus narrower, slightly broader at base, anterior margin straight and expanded in front of compound eyes; eyes weakly projected laterally, interocular distance 1.11 mm; gular sutres indistinct and broadly separated; tentorium with very thin parallel arms. *Antenna (Fig. 96)* : 11 segmented, 1.58 mm long, clavate; scape roughly quadrilateral with 4 long fine setae; pedicel longer than broad, more long than other segments; 3rd to 6th segments more or less quadrilateral and equal; 5th to 8th segments with prominant antero-latal corners; 8th segment transverse, broader than long; 9th segment larger, apically truncate with few fine setae.

Flagellar Formula : SL/W=1.35, PL/W=1.16, 1L/W=2.62, 2L/w=1.5, 5L/W=1.2, 7L/W=0.57, 8L/W=0.54, 9L/W=0.91. *Labrum (Fig. 99)* : 0.41 mm long and 0.56 mm wide, quadrate, anterior corners rounded and covered by setae. *Mandible (Fig. 98)* : 0.42 mm long and 0.38 mm wide, subtriangular with broad base; apical tooth bifid with 0.10 mm depression; basal tooth prominant, slightly curved; membranous prostheca 0.21 mm. *Maxilla (Fig. 97)* : cardo as long as wide, 0.21 mm; stipe 0.37 mm long and 0.18 mm wide with broad base; galea 0.10 mm long, hook like with thick setae; maxillary palpus 4 segmented, 0.78 mm long; 1st segment small 0.06 mm long; 2nd segment 0.14 mm long; 3rd segment small, 0.10 mm long terminal segment 0.48 mm long and 0.42 mm wide, large and securiform. *Labium (Fig. 100)* : post mentum fused with gula; mentum 0.27 mm long and wide, flat; prementum 0.23 mm long and 0.33 mm wide, trapezoidal with straight outer border, anterior corners rounded covered by few fine setae; labial palpus 3 segmented, 0.38 mm long; basal segment small; longer than apical segment; apical segment 0.14 mm long elongated, tapering and bluntly rounded.

Thorax

2.00 mm long and 3.82 mm wide; prothroax broader than long; pronotum 2.00 mm long and 4.22 mm wide, glabrous and without spots, posterior border strongly convex, lateral corners strongly arcuate and marginate; hypomeron slightly depressed; prostermum with carinae, distinctly spereated apically; mesosternum roundly emarginate in the middle of anterior side; metendosternite "Y" shaped with fulcral arms; scutellum triangular, slightly broader and yellowish. *Elytra (Fig. 17)* : 4.22 mm long and 2.20 mm wide, brownish, glabrous with fine punctures; suture not black; base broader and lateral borders not expanded; each elytron with four and half black spots; 2 small band like spots on anterior side; posterior spot transverse, band like, irregular and fuses with its counterpart on another elytron on suture; apical spot small, touching to apex and subrounded; near scuteller region half small spot, fuses with counterpart on another elytron and forms very small somewhat circular spot on suture; elytral epipleurea transversely braod and concave along external margin. *Hind leg (Fig. 101)* : yellowish; coxa 0.23 mm long and 0.71 mm wide, cylindrical; trochanter small, 0.18 mm long and 0.42 mm wide, somewhat squarish and without denticulate process on the inner side; femur 1.26 mm long and 0.50 mm wide, broad, elongated and slightly broader in middle; tibia 1.05 mm long and 0.21 mm wide, elongated narrow at base and gradually broader towards apex with tibial spines on lateral sides; tarsi 4 segmented, 0.84 mm long; 1st segment somewhat conical, 0.42 mm long and 0.14 mm wide, broad at apex, covered by setae on inner and lateral sides; 2nd segment 0.37 mm long and 0.10 mm wide, roundly ended and covered by setae; 3rd segment small, 0.08 mm long and 0.06 wide; terminal segment 0.42 mm long and 0.08 mm wide gradually broad towards apex, terminate into bifid claw; 0.16 mm long with basal tooth.

Abdomen

2.32 mm long and 3.21 mm wide, dorsum yellowish; sternite visible; lateral borders hairy; coxal line on abdominal sternie complete; femoral line incomplete angulate, area around femoral line hairy. *Female genitalia* ovipositor formed by 9th tergite, a pair of 9th pleurite and pair of hemisternite; hemisternite elongated broad at apex with stylus. *Receptaculum seminis (Fig. 102)* : exactly shaped, 0.82 mm long and 0.18 wide, cornu pointed at ti surface with strong markings.

Male

5.12 mm long and 3.85 mm wide, same as female body size. *Male genitialia* : 9th tergite "V" shaped; 10 tergite transverse with coating of setae; apophysis elongate and uniform in width. *Sipho (Fig. 104)* : siphonal capsule well developed; outer lobe quadrate; inner lobe slight curved; siphonal tube broadly curved with digitate apex *Tegmen (Fig. 103)* ; Lateral lobes 0.63 mm long, parallel median lobe, short with broad flat apex coverd by 10 setae; median lobe 0.81 mm long slightly longer than lateral lobes, triangular with broad base, apex narrower and terminate into spatulate rounded structure; tegmen strut elongated, 1.43 mm long and broad at apex.

Holotype

Female, India : Maharashtra, Miraj, coll., 9-X-92, Y.A. Bhosale, on soyabean (*Glycine soja* Linn.), antenna, labrum, mandible, maxilla, labium, hind leg, spermatheca, sipho and tegmen on slides labelled as above.

Paratype

Female 2, Male 2, sex ratio (male : female), 1 : 1 data same as in holopyte. India : Maharashtra, Miraj, coll., from 9-X-92 to 5-I-93, Y.A. Bhosale.

PLATE - 13

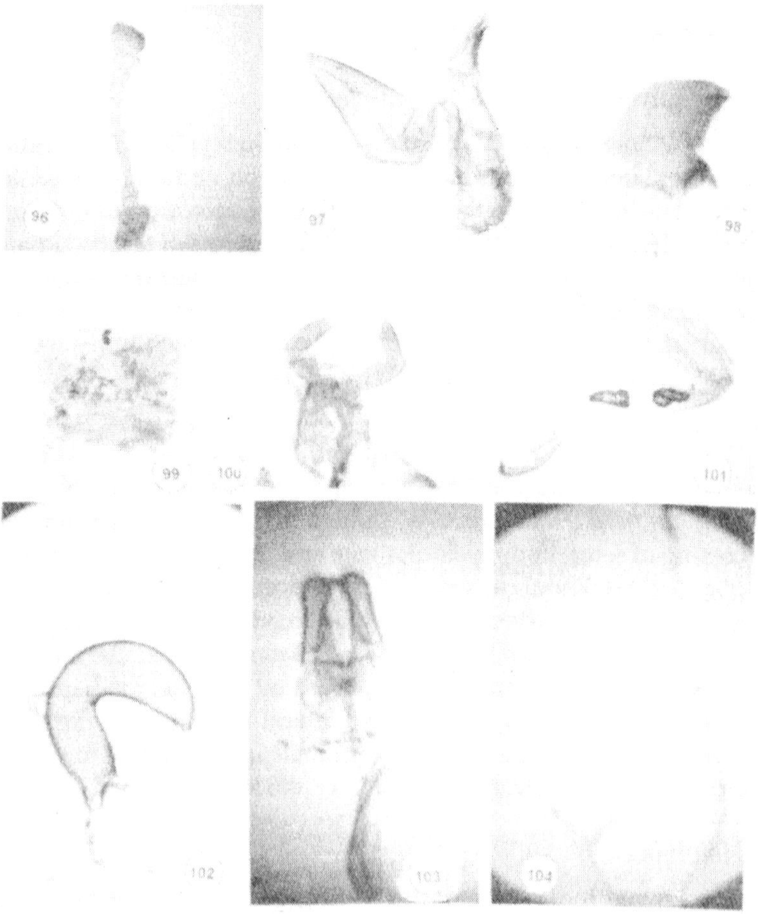

Harmonia soyabinii sp. nov.

Fig. 96 : Antenna

Fig. 97 : Maxilla

Fig. 98 : Mandible

Fig. 99 : Labrum

Fig. 100 : Labium

Fig. 101 : Hind leg

Fig. 102 : Spermatheca

Fig. 103 : Tegmen

Fig. 104 : Sipho

Prey

Aphid, *Aphis craccivora* koch.

Prey Plant

Soyabean, *G. soja.*

DISCUSSION

This species resemble with *H. octomaculata* by antenna, labrum, coxal and femoral lines, it also resembles with *H. axyridis* by the prosternal carinae, pronotum with sub angulate anterior corners. However, it differs from the above species by the following characters :

i) Pronotum without spot,

ii) Terminal segment of maxillary palpus more turncate apically,

iii) Elytra glabrous with transverse band like spot on posterior side which joined on suture with its counterpart,

iv) Area around femoral line on 1st abdominal sternie hairy,

v) Spermatheca exactly "V" shaped, tip of cornu pointed,

vi) Median lobe of tegmen terminate into spatulate rounded structure and

vii) Flagellar formula.

Genus *Menochilus* Timberlake, 1943

Menochilus : Timberlake, 1943, *Hawaii. Plant. Rec.* 47(1):40

Chilomenes : Mulsant, 1850, *Trim-spec. securipalo* : 429.

Chilomenes : Crotch, 1874. *Rev. cocci.* : 179 (pars).

Type species : *Coccinella 6. meaculata* Fabricius, 1781.

The genus *Menochilus* is erected by Timber lake in 1943 which contains only single species, *M. sexmaculatus* described by Fabricius in 1781. The genus is characterised by : body round, slightly longer than wide and medium sized; prosternum with two carinae; anterior margin of mesosternum bisinuate and weakly emerginate in middle; elytral base broader, external margin of elytra not expanded but inclined; 1st abdominal sternum with an oblique lateral line; middle and hind tibia with two spurs.

Menochilus sexmaculatus (Fabricious) 1781

(Plate-2, Fig. 22; Plate-14, Figs. 105-111)

Female

5.20 mm long and 4.02 mm wide, body medium sized and convex on dorsal side; pronotum 1.20 mm long and 2.60 mm wide antenna 0.83 mm long; elytra 4.50 mm long and 3.32 mm wide with 3 black spots; spermatheca 0.58 mm long.

Head

1.30 mm long and 1.80 mm wide, rounded; frons yellowish, without pounctures, fronto clypeal suture lacking clypeus anteriorly projected; gular suture distinct and tentorium with parállel arms; interocular distance 0.85 mm. *Antenna (Fig. 105) :* 11 segmented, 0.83 mm long, clavated scape prominant, more wider, large at base and with strongly rounded inner process with setae; pedicel quadrate; 1st four segments equal, 5th segment smaller than 6th segment : 7th 8th and 9th segment forms club, 9th segment apically truncate with setae.

Flagellar Formula : SL/W=1, PL/W=0.8, 1L/W=1.3, 5L/W=1, 6L/W=0.79, 7L/W=0.4, 9L/W=1.

Labrum : 0.22 mm long and 0.50 mm wide, lateral border rounded with setae. *Mandible (Fig. 106) :* 0.42 mm long and 0.51 mm wide, apical tooth bifid with 0.12 mm depression; basal tooth pointed; membranous prostheca 0.12 mm. *Maxilla (Fig. 107) ;* cardo 0.12 mm long and wide, quadrate; stipe 0.23 mm long and 0.19 mm wide, gradually narrowed towards apex; galea 0.18 mm long with coating of setae; lacinia 0.18 mm long, simple, leaf like with setae; maxillary palpus 4 segmented, 0.90 mm long, basal segment small, 0.10 mm long, 2nd segment 0.20 mm long and 0.09 mm wide, 3rd segment 0.18 mm long, terminal segment large, 0.42 mm long, apically truncate and securiform. *Labium (Fig. 108) :* postmentum and gula fused; mentum trapezoidal, 0.23 mm long and 0.18 mm wide, prementum quadrate, 0.27 mm long and wide, anteriorly covered by numerous hairs; labial palus 3 segmented, 0.28 mm long, basal small, 0.03 mm long; 2nd 0.11 mm long and 0.09 mm wide; terminal segment 0.11 mm long and 0.05 mm wide with rounded apex.

Thorax

1.50 mm long and 2.33 mm wide, prothorax broder; pronotum 1.20 mm long and 2.60 mm wide, anterior margin convex, lateral borders rounded, posterior border shows black margin connected to black border on anterior side by middle vertical black strip; prosternum with pair of prosternal carinae; mesosternum anteriorly flat; metasternum with anterior inter coxal process; metendosternite "Y" shaped; scutellum black, triangular. *Elytra (Fig. 22)* : 4.50 mm long and 3.32 mm wide, glabrous, castaceous; elytral base braod with rounded outer margin; suture black; each elytron with 3 black spots, anterior spot inverted "V" shaped; middle spot band like and not touching to inner and outer border; apical spot small, rounded; elytral apipleureon horizontal, inclined towards apex. *Hind. leg (Fig. 109)* : coxa 0.92 mm long and 0.21 mm wide; trochanter 0.22 mm wide with denticulate process; femur 1.38 mm long and 0.42 mm wide, elongated, broad in middle; tibia 1.40 mm long and 0.25 mm wide, cylindrical with spines and 2 tibial spurs 0.07 mm; tarsi 3 segmented triangular, 0.43 mm long covered by setae, 2nd segment as long as 1st, gradually broad towards apex with dense setae, 3rd segment 0.50 mm long, elongated and terminate into claw; 0.18 mm long curved, bifid with basal tooth.

Abdomen

2.59 mm long and 3.00 mm wide; ventrum yellowish, hairy, lateral borders with setae; six sternites visible, coxal line straight, femoral line complete. *Female genitalia* : 9th tergite, pair of 9th pleurite and a pair of hemisternite together form ovipositer; hemisternite with stylus. *Recepatculum seminis* : 0.58 mm long, curved with weak marking on surface, cornu 0.19 mm long, ramus 0.09 mm long.

PLATE - 14

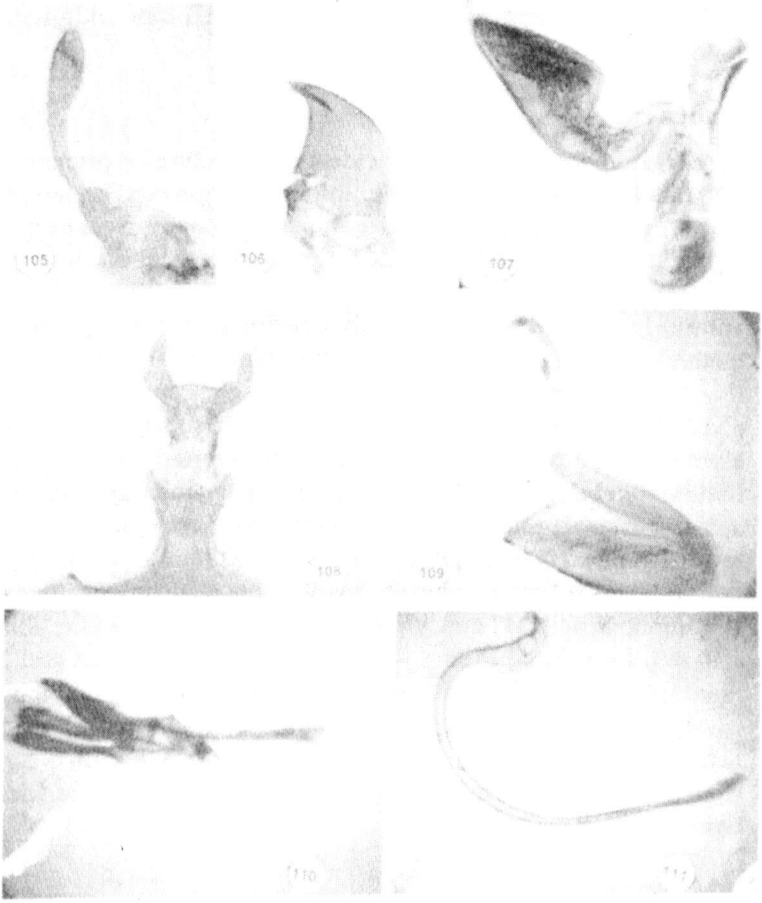

Menochilus sexmaculatus **sp. nov.**

Fig. 105 : Antenna

Fig. 106 : Mandible

Fig. 107 : Maxilla

Fig. 108 : Labium

Fig. 109 : Hind leg

Fig. 110 : Tegmen

Fig. 111 : Sipho

Male

4.61 mm long and 3.70 mm wide, similar like female except body size. *Male genitalia* : 9th tergite 0.19 mm long and 1.10 mm wide, "U" shaped; 10th tergite 0.10 mm long and 0.75 mm wide and strongly transverse with seteae; apophysis 1.00 mm long with rounded apex. *sipho (Fig. 111)* : 2.79 mm long; siphonal capsule well developed, 0.13 mm long and 0.42 mm wide; outer lobe 0.12 mm long and 0.23 mm wide, short, curved; siphonal tube 2.66 mm long, curved. *Tegmen (Fig. 110)* : lateral lobes as long as median lobe, 0.70 mm, runs parallel, apex covered by setae; median lobe triangular with blunt apex; tegmen strut 1.20 mm long, expanded at apex.

Holotype

Female, India : Maharashtra, Miraj, coll. 10-VII-92, Y.A. Bhosale, on Jowar (*Sorghu vulgare* Pers.), antenna, labrum, mandible, maxilla, labium, spermatheca, sipho and tegmen of male on slides labelled as above.

Paratype

Female 90, male 75, sex ratio (Male: Female) 1:1.2 data same as in holotype coll. for 10-VII-92 to 25-II-94, Y.A. Bhosale, India Maharashtra, Miraj.

Prey

Aphid, *Rhopalosiphum maidis* (Fitch).

Prey Plant

Jowar, *S. vulgare.*

Menochilus tuljapurensis sp. nov.

(Plate-2, Fig. 21; Plate-15, Figs. 112-116).

Female

4.64 mm long and 3.53mm wide, body oval; pronotum 1.47 mm long and 2.80 mm wide; antenna 0.82 mm long; elytra 3.82 mm long and 2.32 mm wide with 2 black spots; spermatheca 0.60 mm long.

Head

1.20 mm long and 1.64 mm wide, rounded; frons black with "V" shaped margin, rough; clypeus yellow, anteriorly projected; eyes black, shiny, laterally projected, interocular distance 0.82 mm. *Antenna (Fig. 112)* : 11 segments, 0.82 mm long clavate; scape more broader than long, inner process rounded and produced; pedicel quadrate; 1st to 4th segments equal ; 6th longer than 5th segment; terminal 3 segment form compact club, terminal segment apically truncate and covered by setae.

Flagellar formula : SL/W=1, PL/W=0.7, 1L/W=1.3, 5L/W=1, 6L/W=0.85, 7L/W=0.7, 9L/W=1.

Labrum (Fig. 115) : 0.29 mm long and 0.50 mm wide, quadrate, lateral borders rounded, yellowish. *Mandible (fig. 114)* : 0.42 mm long and 0.49 mm wide, triangular, apical tooth bifid with 0.15 mm depresion ; basal tooth curved; membranous prostheca 0.19 mm. *Maxilla (Fig. 116)* : cardo 0.20 mm long and wide, quadrate; stipe 0.29 mm long and 0.17 mm wide, gradually broad towards apex; galea 0.12 mm long with setae; lacinia as long as galea leaf like with setae; maxillary palpus 4 segmented, 1.00 mm long, basal segment small, 0.14 mm long; 2nd segment 0.21 mm long and 0.10 mm wide; 3rd segment 0.19 mm long, as wide as 2nd segment; terminal segment large, 0.43 mm long, broad at apex, apically truncate and securiform. *Labium (Fig. 113)* : post mentum and gula fused; mentum triangular, 0.25 mm long and 0.18 mm wide; prementum quadrate, 0.27 mm long and wide with hairs on anterior border; labial palpus 3 segmented, 0.29 mm long, basal segment 0.06 mm long; 2nd segment 0.10 mm long and 0.08 mm wide; terminal segment 0.12 mm long and 0.06 mm wide, and roundly ended.

Throax

1.89 mm long and 2.84 mm wide; pronotum 1.47 mm long and 2.80 mm wide, yellowish, glabrous, anteriorly convex in middle, lateral borders rounded, posterior border black connected to anterior black patch by vertical black strip; anteriorly mesosternum excavated, metendosternite with paired lateral arms, scutellum black. *Elytra (Fig. 21)* : 3.82 mm long and 2.32 mm wide, glabrous, very peculiar finely punctate, base slightly depressed in middle, outer margin strongly inclined posteriorly, apex trianglar and pointed, inner border black, suture black, each elytron with 2 black spots, basal spot some what quadrate, hardly touching to basal border, posterior border of spot in middle shows depression; 2nd spot large completely occupies posterior part also covers apical tip, anterior border of spot somewhat "W" shaped with middle, transverse, reaching upto apex. *Hind leg* : coxa 0.90 mm long and 0.21 mm wide; trochanter 0.29 mm wide with denticulate process; femur 1.40 mm long and 0.42 mm wide elongated; tibia 1.48 mm long and 0.29 mm wide with tibial spines and with 2 tibial spurs 0.07 mm; tarsi 3 segmented, 1st segment 0.42 mm rounded at apex; 3rd segment 0.52 mm long with setae and terminate into claw; claw 0.18 mm long, bifid with basal tooth.

Abdomen

2.53 mm long and 3.37 mm wide, dorsum black, ventrum hairy, lateral margins with setae; six sternite visible; coxal line complete; femoral line complete and slightly angulate. *Female genitalia* : hemisternites alongated with stylus. *Receptaculum seminis* : 0.60 mm long, curved with weak markings on surface, broad at base, cornu 0.17 mm and ramus 0.09 mm long.

Male

Unknown.

Holotype

Female, India : Maharashtra, Tuljapur, coll. 14-XII-93, Y.A. Bhosale on Sunflower (*Helianthus annus* Linn.), antenna, labrum, mandible maxilla and labium on slides labelled as above.

Paratype

Female 3, data same as in holotype coll. 14-XII-93 Y.A. Bhosale, India : Maharashtra Tuljapur.

Prey

PLATE - 15

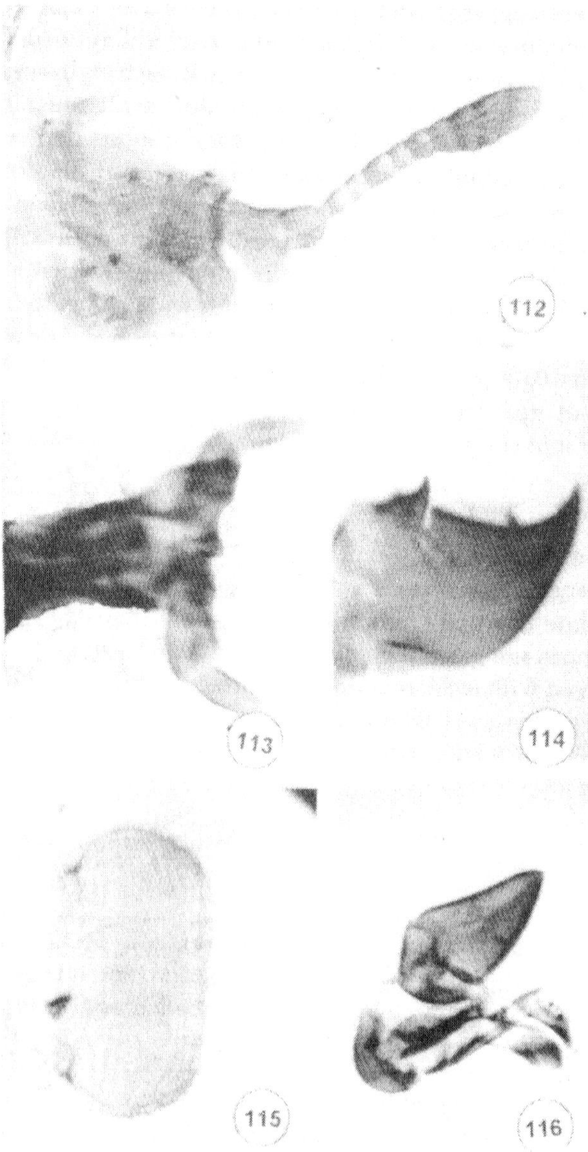

***Menochilus tuljapurensis* sp. nov.**

Fig. 112 : Antenna
Fig. 113 : Labium
Fig. 114 : Mandible
Fig. 115 : Labrum
Fig. 116 : Maxilla

Prey Plant

Sunflower, *H. annus.*

DISCUSSION

This species runs close to *M. sexmaculatus* having antenna, maxilla, mandible, labium pronotum, scutellum, spermatheca. But it differ from the above species by having the following characters :

i) At once distinguished by elytral spot umbrella shaped,

ii) Anterior margin of frons black,

iii) Femoral line complete and slightly angulate and

iv) Flagellar formula.

Menochilus shivajensis sp. nov.

(Plate-2, Fig. 23; Plate-16, Figs, 117-125)

Female

5.48 mm long and 4.65 mm wide, body round, medium sized; pronotum 1.00 mm long and 2.53 mm wide; antenna 0.75 mm long; elytra testaceous, 4,43 mm long and 3.16 mm wide with 3 black spots; sprematheca 0.63 mm long with "W" shaped inner part.

Head

1.26 mm long and 1.75 mm wide, somewhat rounded with black border on posterior side; frons yellowish with half star like margins on posterior side, smooth and without punctures, fronto clypeal suture indistinct; clypeus anteriorly excavated and slightly expanded infornt of compound eyes; gular sutures slightly distinct; tenorium with tentorial arms; compund eyes laterally placed, inter ocular distance 0.84 mm. *Antenna (Fig. 117)* : 11 segmented, 0.75 mm long, clavate with compact club; scape large, broad at apex and quadrate at base, more broader than long inner process strongly ronded and prodcued with 3 to 4 setae; pedicel some what quadrate and with single setae; 1st to 4th segment of flagellum more or less equal; 5th segment small; 6th slightl longer than 5th; last 3 segments distinctly forms club, 7th and 8 the segment more transverse, 9th segment slightly longer than width, apically truncate and covered by long fine setae.

Flagellar formula : SL/W=1, PL/W=0.6, 1L/W=1.5, 5L/W=1, 6L/W=0.75, 7L/W=0.6, 9L/W=1

Labrum (Fig. 121) : 0.27 mm long and 0.56 mm wide, squarish plate like with rounded borders, yellowish with setae on anterior and lateral border. *Mandible (Fig. 118)* : 0.39 mm long and 0.48 mm wide, apical tooth bifid with 0.10 mm depression; basal tooth pointed; membranous prostheca 0.15 mm. *Maxilla (Fig. 119)* ; cardo 0.23 mm long and wide, rounded, squarish; stipe 0.27 mm long and 0.18 mm wide, broad at base and narrow towards apex; galea 0.16 mm long, hood like with dense coating of setae; lacinia 0.16 mm long, simple, leaf like with coating of setae; maxillary palpus 4 segmented, 0.94 mm long, basal segment small, 0.12 mm long; 2nd segment 0.21 mm long and 0.1 mm wide, as wide as 3rd segment; 3rd segment 0.18 mm long with single seta; terminal segment large, 0.42 mm long and wide, broad towards apex, apically truncate securiform. *Labium (Fig. 120)* : postmentum fused with gula; mentum trapezoidal, 0.21 mm long and 0.16 mm wide; prementum subquadrate, 0.25 mm long and wide, apically bordenes by numerous hairs; labial palpus 3 segmented, 0.25 mm long, basal segment very small, 0.04 mm long; 2nd segment 0.10 mm long and 0.08 mm wide; terminal segment 0.10 mm long and 0.04 mm wide, small and bluntly pointed.

Thorax

1.47 mm long and 2.21 mm wide; prothorax border than long; pronotum 1.00 mm long and 2.53 mm wide, anterior margin strongly excavated, lateral sides rounded, smooth without punctures, black borders on posterior side, connected to black border on anterior side through vertical black strip in middle; hypomeron hardly foveate; prosternum with two carinae anterior margin of mesosternum flat; metasternum slightly triangular with anterior inter coxal process, lateral margins slightly convergent anteriorly; metendosternite "Y" shaped with stalk and paired arms of fulcra; scutellum black, triangular. *Elytra (Fig. 23)* : 4.43 mm long and 3.16 mm wide. tesaceous, glabrous and sparsely punctured;

elytral base braod and more inclined below, outer rounded, suture black; each elytron with 3 black spots, anterior spot inverted "V" shaped; middle spot band like, irregular and not touching to inner and outer border of elytron; apical spot small, some what rounded; elytral epipleuron entire and strongly inclined towards apex. *Hind leg (Fig. 122)* : coxa 0.99 mm long and 0.27 mm wide; trochanter 0.37 mm wide, triangular with denticulate process on inner side; femur 1.32 mm long and 0.40 mm wide, broad elongated; tibia 1.39 mm long and 0.21 mm wide, cylindrical with tibial spines more concentrated towards the apex and with 2 tibial spurs 0.08 mm; tarsi 3 segmented triangular, 0.40 mm long, densely covered by setae, 2nd segment 0.40 mm long, narrow at base and rounded towards apex with dense coating of setae; 3rd segment 0.54 mm long, elongated, gradually broad towards apex with few setae on surface and terminate into claw 0.16 mm long bifid, curved with basal tooth.

Abdomen

2.63 mm long and 3.05 mm wide, dorsum black in middle; ventrum yellowish pubescent; lateral borders covered by setae; six sternite visible; coxal line straight; femoral line complete slightly angulate. *Female genitalia* : hemisternite elongated with stylus. *Receptaculum seminis (Fig. 125)* : 0.63 mm long, cylindrical, curved with "W" loke inner side, weak markings on surface, broad at base, cornu 0.21 mm long, rounded ramus 0.10 mm long.

Male

4.64 mm long and 3.79 mm wide, similar like female except body size. *Male genitalia* : 9th tergite 0.21 mm long and 1.16 mm wide, "U" shaped; 10th tergite 0.10 mm long and 0.80 mm wide, transverse and covered by setae; apophysis 1.05 mm long and terminate roundly. *Sipho (Fig. 124)* : 2.99 mm long; siphonal capsule well developed, 0.14 mm long and 0.42 mm wide, outer lobe 0.12 mm long and 0.25 mm wide, elongated quadrate, inner lobe 0.08 mm long and 0.16 mm wide, short, slight broad, curved and roundly ended, notch between two lobes prominent; siphonal tube 2.84 mm long, immediately curved, of tube very fine, more longer and shows coiling. *Tegmen (Fig. 123)* : lateral lobes as long as median lobe, 0.73 mm , parallel and terminate roundly, apex covered by setae; middle lobe flat, some what triangular with blunty pointed apex tegmen strut 1.22 mm long, elongated and expanded towards.

Holotype

Female, India : Maharashtra, Miraj, coll., 25-XII-93, Y.A. Bhosale, on Jowar (*Sorghum vulgare* Pers.), antenna, labrum, maxilla, labium, 1st abdominal sternite, spermatheca, sipho, last terga of male and tegmen on slides labelled as above.

Paratype

Female 24, male 12, sex ratio (Male : Female), 1:2 data same as in holotype, coll., from 25-XII-93 to 9-III-94, Y.A. Bosale, on Maize (*Zea mays* Linn.) ; Jowar (*S. vulgare*) and Hybrid jowar CSH-5, India : Maharashtra, Miraj.

Prey

Aphid, *rhopalosiphum maidis* (Fitch).

Prey Plant

Jowar, *S. vulgare,* Maize, *Z. mays,* Hybrid Jower CSH-5.

DISCUSSION

This species resembles to *M. sexmaculatus* by having antenna 11 segmented, clavate, maxilla, labium scutellum and same elytral spots. However, it differs from the above species hy having the following characters :

i) Elytra testaceous, glabrous and sparsely punctured,

ii) Anterior margin of frons black and half star shaped,

iii) Femoral line complete and reaches to the posterolateral part of sternum,

iv) Siphonal apex thread like and more elongated, notch between siphonal lobes very prominent,

PLATE - 16

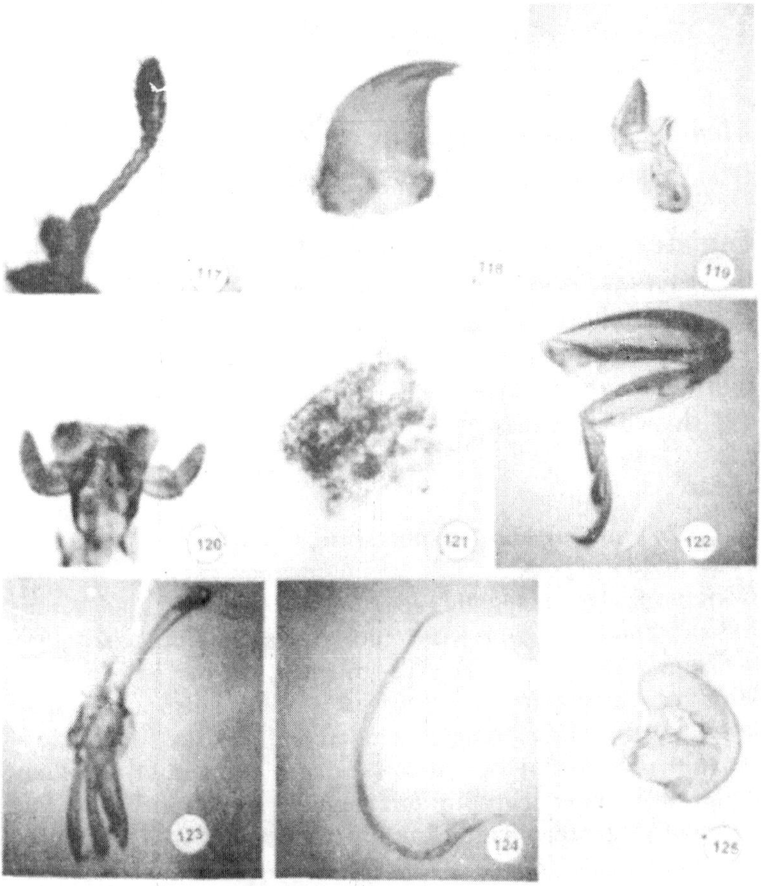

Menochilus shivajensis **sp. nov.**

Fig. 117 : Antenna

Fig. 118 : Mandible

Fig. 119 : Maxilla

Fig. 120 : Labium

Fig. 121 : Labrum

Fig. 122 : Hind leg

Fig. 123 : Tegmen

Fig. 124 : Sipho

Fig. 125 : Spermatheca

v) Median lobe of tegmen triangular with bluntly pointed apex.

vi) Inner part of spermatheca "W" shapped and

vii) Flagellar formula.

Menochilus kharipi sp. nov.

(Plate-2, Fig. 24; Plate-17, Figs, 126-132)

Female

5.00 mm long and 4.00 mm wide, body medium sized, elongated oval and convex on dorsal side, dorsum black; pronotum 1.68 mm long 3.16 mm wide; antenna 0.68 mm long; elytra black 4.00 mm long and 3.00 mm wide; spermatheca 0.60 mm long, with "V" shaped inner part.

Head

1.26 mm long and 1.66 mm wide, rounded with depression in the middle of posterior side, lateral sides covered by dense setae; frons black; clypeus slightly expanded and yellowish; gular suture indistinct and tentorium with tentorial arms. interocular distance 0.97 mm. *Antenna (Fig. 126)* : 11 segmented, 0.68 mm long, clavate, last three segments from compact club; scape squarish at base, apex large, inner surface strongly rounded and produced with 2 to 3 setae; pedicel with single seta; 1st to 4th segment of flagellum equal; 6th longer than 5th; 8th segment more transverse; 9th segment longer than width, apically truncate and with long setae.

Flagellar formula : SL/W=1, PL/W=0.5, 1L/W=1.2, 6L/W=0.60, 8L/W=0.6, 9L/W=1.

Labrum (Fig. 130) : 0.23 mm long and 0.50 mm wide, plate like, lateral borders rounded. *Mandible (Fig. 127)* : 0.45 mm long and 0.53 mm wide, triangular, apical tooth bifid with depression 0.10 mm; basal tooth curved and pointed; membranous prostheca 0.18 mm. *Maxilla (Fig. 128)* : cardo 0.20 mm long and wide, gradually narrowed towards apex; galea 0.12 mm long, triangular with setae; maxillary palpus 4 segmanted, 0.9 mm long, basal segment small, 0.10 mm long, 2nd 0.19 mm long and 0.10 mm wide, 3rd segment

0.18 mm long; terminal segment large, 0.33 mm long and wide, apically truncate, securiform. *Labium (Fig. 129) :* post mentum and gula fused; mentum triangular, 0.2 mm long and 0.16 mm wide, prementum roughly quadrate, palpus 3 segmented, 0.21 mm long basal segment 0.02 mm long; 2nd 0.12 mm long and 0.07 mm wide; terminal segment 0.12 mm long and 0.03 mm wide, bluntly pointed.

Thorax

1.62 mm long and 2.45 mm wide; pronotum 1.6 mm long and 3.16 wide, glabrous finely punctate with middle black vertical marking connected with inner large black quadrate patch, anterior border convex in middle with setae, antero lateal borders strongly angulate, postero lateral borders rouned, posterior black marking connected by vertical black marking with anterior quadrate black patch, hypomeron foveate; prosternal carinae runs parallel scutellum black. *Elytra (Fig. 24) :* 4.00 mm long and 3.00 mm wide, black, glabrous, finely punctured outer border inclined below, elytron without spots; elytral epipleuro reaches to apex. *Hind Leg (Fig. 131) ;* coxa 0.82 mm long an 0.22 mm wide; trochanter 0.29 mm with denticulate process; femur 1.42 mm long and 0.38 mm wide, elongated; tibia 1.4 mm long and 0.25 mm wide with tibial spines and tibial spurs 0.07 mm; tarsi 3 segmented, 1st segment conical, 0.3 mm long covered by setae; 2nd segment 0.38 mm long with rounded apex covered by setae; 3rd segment 0.60 mm long, gradually widen towards apex and terminate into claw; claw 0.18 mm long, bifid, curved with tooth a base.

Abdomen

2.6 mm long and 3.20 mm wide, dorsum black, ventrum pubescent; lateral borders with setae; six sternite visible; coxal line complete; femoral line complete and reaching to outer border. *Female genitalia ;* hemisternite elongated with stylus. *Receptaculum seminis (Fig. 132) :* 0.75 mm long, curved, "C" shaped with "V" like inner margin, close markings on surface, broad at base, cornu 0.22 mm long, ramus 0.15 mm long.

PLATE - 17

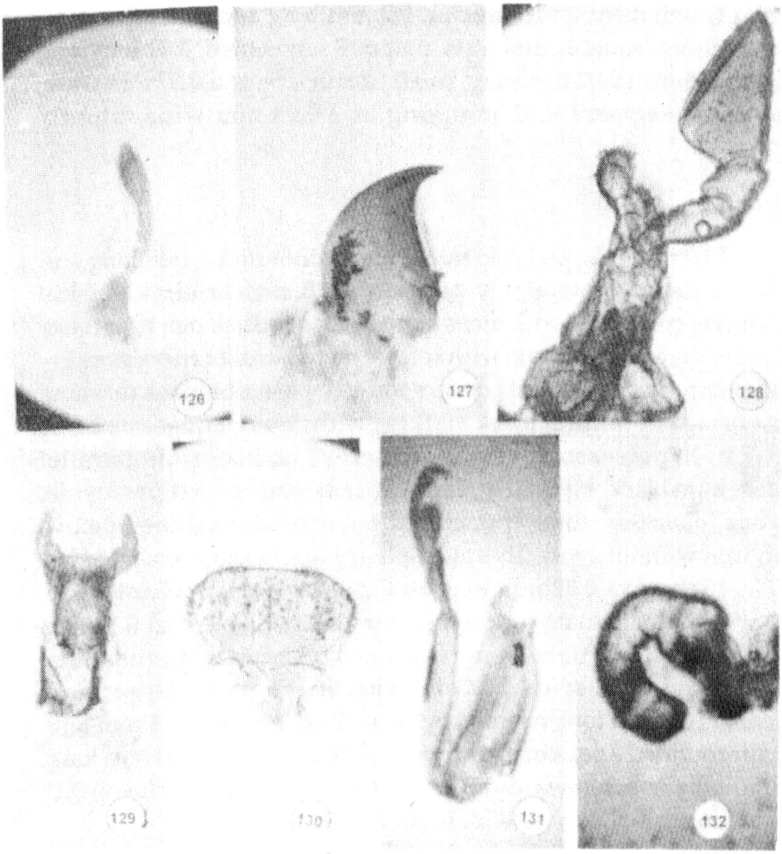

Menochilus kharipi sp. nov.

Fig. 126 : Antenna

Fig. 127 : Mandible

Fig. 128 : Maxilla

Fig. 129 : Labium

Fig. 130 : Labrum

Fig. 131 : Hind leg

Fig. 132 : Spermatheca

Male

Unknown.

Holotype

Female, India : Maharashtra, Miraj, coll. 23-I-94, Y.A. Bhosale, on Maize (*Zea mays* Linn.) antenna, labrum, mandible, maxilla, labium and sprematheca on slides labelled as above.

Paratype

Female 2, data same as in holotype, coll. 23-I-94 Y.A. Bhosale on Maize (z. *mays*), India Maharashtra, Miraj.

Prey

Aphid, *Rhopalosiphium maidis* (Fitch).

Prey Plant

Maize, Z. *mays*.

DISSCUSSION

This species runs close to *M. sexmaculatus* by haivng antenna, mandible, maxilla, pronotum. However, it differs from the above species by having the following characters :

i) Body large in size,

ii) Elytra black and without spots,

iii) Femoral line complete not angulate and reaches to posterior border,

iv) Sprematheca with "V" like inner part and

v) Flagellar formula.

Genus *Verina* Milsant, 1850

Coccinella discolor Fabricius, 1798, *Suppl. Ent. Syst.* : 77 *Vernia discolor* : Mulsant, 1850, *Spec. Irim. securipalp* : 369 Type species : *vernia discolor.*

The genus is proposed by Muslant in 1850 and it includes 3 species *viz; V. discolor, V. kiotoensis* and *V. cardoni*. Omkar and Rind

(1992) reported *V. cardoni* from Uttar Pradesh. The genus is characterised by : body hemispherical, glabrous; dorsum orange brown; frons sometimes withblack spot; pronotum with discal spots; scutellum black; elytron with very narrow black sutural area; femora brown to black; tibia and tarsi brown; prosternal carinae reaching half of prosternal length.

Vernia polyphagi sp. nov.

(Plate-2, Fig. 28; Plate-18, Figs. 133-140)

Male

Medium sized, 4.02 mm long and 3.87 mm wide, oval, dorsum convex, underside black; pronotum 1.02 mm long and 2.89 mm wide; antenna 0.94 mm long, 11 segmented; elytra reddish brown, 3.0 mm long and 1.83 mm wide; sipho with deep notch between lobes of siphonal capsule.

Head

1.47 mm long and 1.68 mm wide, quadrate, inserted into pronotum with red brown apices of mandibles; frons without black spot, fronto clypeal suture indistinct; clypeus broad expanded infront of compound eyes; eyes laterally projected with 0.85 mm interocular distance. *Antenna (Fig. 133) :* 11 segmented, 0.94 mm long; scape large, quadrate with setae, it slightly produced inwardly; pedicel elongated narrower with setae; 1st to 5th segments more or less similar and each with single seta; 6th small; last 3 segments somewhat large; 8th segment elongated large with numerous setae.

Flageilar formula : SL/W=1, PL/W=0.35, 1L/W=0.25, 5L/W= 0.35, 7L/W=0.75, 8L/W=0.83, 9L/W=1.

Labrum (Fig. 136) : 0.31 mm long and 0.61 mm wide, somewhat oval with rounded lateral borders, surface and anterior border covered by setae. *Mandible (Fig. 134) :* 0.54 mm long and 0.34 mm wide, triangular with bifid apical tooth and depression between bifid tooth 0.10 mm; basal tooth single, curved and pointed; membranous prostheca 0.23 mm. *Maxilla (Fig. 135) :* cardo as long as wide, 0.29 mm long and wide; stipe longer than wide and as wide as cardo; galea spatulate, 0.23 mm long and covered by setae; lacinia as long as galea, simple, elongated with setae on inner side; maxillary palpus 4 segmented; 1st segment very small, 0.08 mm; 2nd smaller than 3rd segment; 3rd segment 0.91 mm long; terminal segment large, 0.31 mm long and 0.29 mm wide, apically truncate and securiform. *Labium* : gula and submentum fused; mentum small quadrate; 0.19 mm long and wide; prementum trapezoidal, apically broad, 0.20 mm long and 0.28 mm wide; labial palpus 3 segmented; 1st segment very small, 0.04 mm; 2nd segment tapering, roundly ended, 0.18 mm long.

Thorax

1.09 mm long and 2.98 mm wide; pronotum yellowish red, 1.02 mm long and 2.89 mm wide, lateral sides rounded, densely punctured, basal border with black marking joins with anterior black marking by vertical black strip; prosternal carinae half; mesosternum broader than long; metendosternite with pair of fulcral arms, scutellum black. *Elytra (Fig. 128)* : 3.0 mm long and 1.93 mm wide, reddish brown, glabrous, distinctly sparsely punctured; inner border black, suture black, elytron without spots, basal border slightly wider than pronotum outer margin inclined more towards apex. *Hind leg (Fig. 138)* : coxa strong and stout, 1.32 mm long and 0.89 mm wide; trochanter small with denticulate process; femur elongated, 1.89 mm long and 1.2 mm wide, slightly broad in middle; tibia brown more elongated than femur, 1.94 mm long with setae and tibial spines, tibal spurs not prominent; tarsi brown, 3 segmented; 1st segment triangular broad towards apex and laterally covered by setae, 0.83 mm long, 2nd segment elongated with setae on lateral side, 0.90 mm long; terminal segment elongated, gradually broad toward apex, 1.20 mm long and terminate into bifid claw; claw curved, 0.12 mm, pointed with basal tooth.

Abdomen

2.00 mm long and 2.99 mm wide, dorsum black with yellow border; six sternites visible; laterally covered by setae; ventrum hairy; coxal line on 1st abdominal sternite complete; femoral line slightly angulate and complete, area around femoral line hairy.

Male genitalia : 9th tergite "U" shaped; 10th tergite strongly transverse and covered by setae; apophysis elongated, similar in width and without bifid base (Fig. 137). *Sipho (Fig. 140)* : siphonal capsule well developed; inner lobe very prominant; siphonal tube immediately curved, apex of tube pointed and upwardly directed. *Tegmen (Fig. 139)* : lateral lobes as long as median lobe, 0.85 mm densely covered by setae at apex, apex slightly swollen; median lobe boat shaped with narrow apex; tegmen stout elongated broad at apex.

Female

Unknown.

Holotype

Male, India : Mahrasthra, Miraj Coll., 5-IX-92, Y.A.Bhosale, on Hybrid jowar CSH-5, antenna, labrum, mandible, maxilla, labrum, sipho and tegmen on slides lebelled as above.

Paratype

Male 4, data same as in holotype, coll., 5-IX-92, Y.A. Bhosale, on hybrid jowar CSH-5, India : Maharashtra, Miraj.

Prey

Aphid, *Rhopalosiphum maidis* (Fitch).

Prey Plant

Hybrid Jowar, CSH-5.

DISCUSSION

This species runs close to *V. discolor* and *V. kiotoensis* by having following characters : antenna, labrum, mandible, mexilla, pronotum and scutellum. But it differs from above species by having following characters :

i) Elytra reddish brown and sparsely punctured.

ii) Pronotum with basal and apical black borders.

PLATE - 18

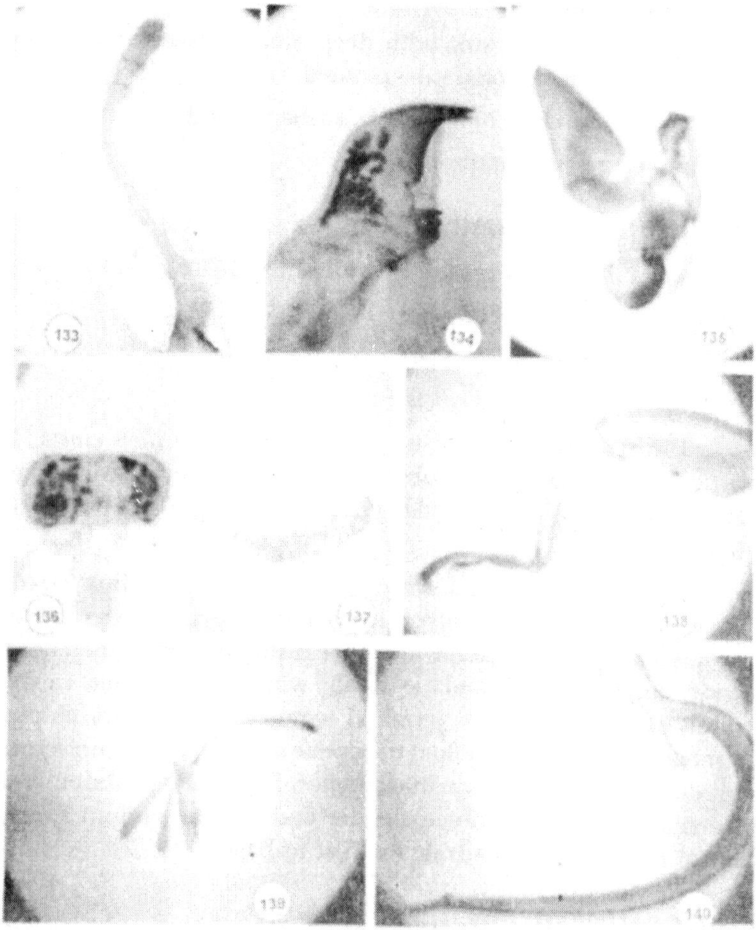

Vernia polyphagi sp. nov.

Fig. 133 : Antenna

Fig. 134 : Mandible

Fig. 135 : Maxilia

Fig. 136 : Labrum

Fig. 137 : Last terga of male

Fig. 138 : Hind leg

Fig. 139 : Tegmen

Fig. 140 : Sipho

iii) Femoral line on 1st abdominal sternite complete and area around femoral line hairy.

iv) Siphonal capsule with deep notch between lobes and apex of siphonal tube pointed.

v) Median lobe of tegmen boat shaped and

vi) Flagellar formula.

Genus *Illeis* Mulsant, 1850

The genus *Illeis* was errected by Mulsant in 1850.

Illeis Mulsant, 1850, *Securipalp* : 1026

Egleis Mulsant, 1850 *I. C.* : 167

Type species : *Coccinella cincta* Fabricius, 1798.

The species described under this genus are *Illeis cincta, I. bistigmosa* and *I. indica* and subspecies *I. cincta amamiana* (Miyatake, 1959; Crotch, 1874; Korschefsky, 1932, Timberlake, 1943; Hielawski, 1961).

The genus shows following characters : body medium sized, short, oval and weekly convex; frons narrow; eyes with a shallow postantennal emargination; antennae distinctly longer than head width and very thin; mandible usually with multidenticulate apex, rarely with a bifid apex; terminal segment of maxillary palpus strongly transverse; scutellum transverse and tirangular; prosternal process with parallel carinae; femoral line of 1st abdominal sternum incomplete; tibiae slender and simple without spurs; tarsal claws with a quadrate tooth at its base.

Illeis sathei sp. nov.

(Plate-2, Fig. 19; Plate-19, Figs. 141-148)

Male

Body short, 4.85 mm long and 3.26 mm wide, slightly elongated whitish yellow; pronotum 1.05 mm long and 2.32 mm wide, without spots; antenna 1.37 mm long; elytra 3.79 mm long and 2.42 mm wide; middle and apex of lateral lobes of tegmen covered by setae.

Head

Black, 1.35 mm long and 1.26 mm wide, somewhat quadrate with slightly projected anterior side; frons pale, fronto clypeal suture indistinct; clypeus expanded in front of compound eyes; mouth parts yellowish brown; compound eyes black, laterally projected, interocular distance 0.42 mm; dorsum smooth; lateral borders covered by setae; ventrum smooth, black; gular sutures prominant and run parallel; tentorium with tentorial arms. *Antenna (Fig. 141) :* 11 segmented, 1.37 mm long, clavate and somewhat serrated; scape large, more longer than wide with six fine setae on inner side; pedicel as broad as long, somewhat quadrate with 2 fine setae on inner side; 1st, 2nd and 3rd segments longer than other segments each with one fine seta on inner side; 4th, 5th, and 6th segments more or less similar, 5th segment with one long seta and 5 fine setae on inner anterolateral side; 7th, 8th and 9th segments larger than preceeding segments and forms distinct club, 9th segment larger than other, narrow at base and apically broader with 6 long setae; yellowish with brown tinge on scape and terminal segment.

Flagellar formula : SL/W=1.5, PL/W=1.5, 1L/W=2.3, 3L/W=2.6, 5L/W=1.5, 7L/W=1, 8L/W=1, 9L/W=1.16.

Larbum : oval, yellowish, sub rounded, lateral sides covered by setae with elevations in middle. *Mandible (Fig. 142) :* 0.42 mm long and 0.34 mm wide at base, somewhat triangular with broad base; apical tooth bifid, brownish, sharply pointed with depression 0.12 mm; basal tooth small, very slightly curved and abruptly pointed; membranous prostheca 0.31 mm. *Maxilla (Fig. 143) :* cardo as broad as long, 0.23 mm; stipe longer than width, 033 mm long and 0.23 mm wide; galea large, hood like, 0.25 mm long with thick dense setae; lacinia 0.12 mm long, small, simple, leaf like with fine setae on inner side; maxillary palpus 4 segmented, 0.82 mm long; basal segment small, 0.06 mm long and wide; 2nd mm segment 0.29 mm long and 0.12 mm wide, narrow at base and wide, small, quadrate; terminal segment large, 0.31 mm long and 0.61 mm wide, more transverse, capitulate, narrow at base and apically expanded. *Labium (Fig. 144) :* post mentum fused with gula; mentum broad, 0.18 mm long and 0.16 mm wide; prementum triangular, 0.23 mm long and 0.46 mm wide, apically broad and not covered by hairs; labial palpus 3 segmented, 0.31 mm long, 1st segment very small; 2nd as long as broad; 3rd segment bluntly pointed.

Thorax

1.79 mm long and 2.57 mm wide; prothorax broader than long; pronotum 1.05 mm long and 2.32 mm wide, quadrate, pale yellow, smooth, without spots and with rounded lateral borders; hypomerone slightly foveate; prosternum with parallel carinae, not reaching to anterior border of sternum; anterior margin of mesosternum slightly covex; metasternum slightly triangular; matendosternite "Y" shaped with paired arms; scutellum triangular. *Elytra (Fig. 19)* : 3.79 mm long and 2.42 mm wide, pale yellowish without spots, finely punctate in middle and densely punctate on outer margin; when elytra on body, it shows black stipes. elytral epipleuron pale, without punctures, transversly broad and reaching to apex. *Hind leg (Fig. 146)* : caxa 0.88 mm long and 0.31 mm wide, shortly elongated and broad; torchanter somewhat triangular, 0.37 mm long and 0.27 mm wide, cap like structure on femur with prominant denticulate process on inner side; femur elongated, 0.69 mm long and 0.35 mm wide, broad in middle; tibia more elongated, 1.54 mm long and 0.23 mm wide, cylindrical with tibial spines on all sides and without tibial spurs; tarsi 3 segmented, 0.84 mm long, 1st segment elongated, 0.42 mm long with setae on inner side; 2nd segment triangular, 0.46 mm long with setae on anterior border; terminal segment 0.44 mm long, flattened, narrow at base and gradually broad towards apex and terminate into bifid claw; claw 0.16 mm long, pointed with sub quadrate basal tooth.

Abdomen

2.32 mm long and 3.16 mm wide; dorsum yellowish; ventrum pubescent; 8 terga visible, broader at base and narrower towards apex; six sternite visible; lateral border covered by setae; coxal line on 1st abdominal sternite straight; femoral line incomplete and reaching to posterior border of sternum (*Fig. 145*). *Male genitalia* : 9th tergite 0.16 mm long and 1.47 mm wide, sipho "U" shaped;

10th tergite strongly transverse 0.14 mm long and 0.16 mm wide, covered by setae; apophysis 0.78 mm long, elongated, base not bifid. *Sipho (Fig. 148)* : siphonal capsule 0.42 mm long and 0.37 mm wide, well developed, outer lobe quadrate, 0.16 mm long and 0.25 mm wide, inner lobe slightly curved, 0.08 mm long and 0.12 mm wide; siphonal tube immediately curved, 3.05 mm long and 0.08 mm wide uniform in width up to middle and then gradually smaller towards apex, apex curved and terminate into pointed structure with characteristic structure on curvature. *Tegmen (Fig. 147)* : tegmen strut 0.99 mm long, similar in width through out length with slightly borad and flat apex; lateral lobes 0.63 mm long, narrow in midlle and slightly broad rounded at apex with long setae on middle and apex; median lobe 0.73 mm long, narrow towards apex and terminate hook like.

Female

Unknown.

Holotype

Male, India : Maharashtra, Kolhapur coll, 9-IX-93, T.V. Sathe, on Sauf (*Foeniculuns vulgare* Mill), antenna, labrum, mandible, labium, 1st abdominal sternite, last terga of male, sipho and tegmen on slides labelled as above.

Paratype

Male 2, India Maharashtra, Kolhapur, coll., T.V. Sathe, on 10-IX-93, data same as holotype.

Prey

Aphid, *Aphis craccivora* Koch.

Prey Plant

Sauf. *F. vulgare*.

DISCUSSION

This species resembles with *Illeis bistigmosa* by having 11 segmented, serrated antenna, maxillary palpus maxilla and colour of body. It also resembles with *I. indica* by having the median lobe of tegmen slightly longer than lateral lobes. But it differs from the

PLATE - 19

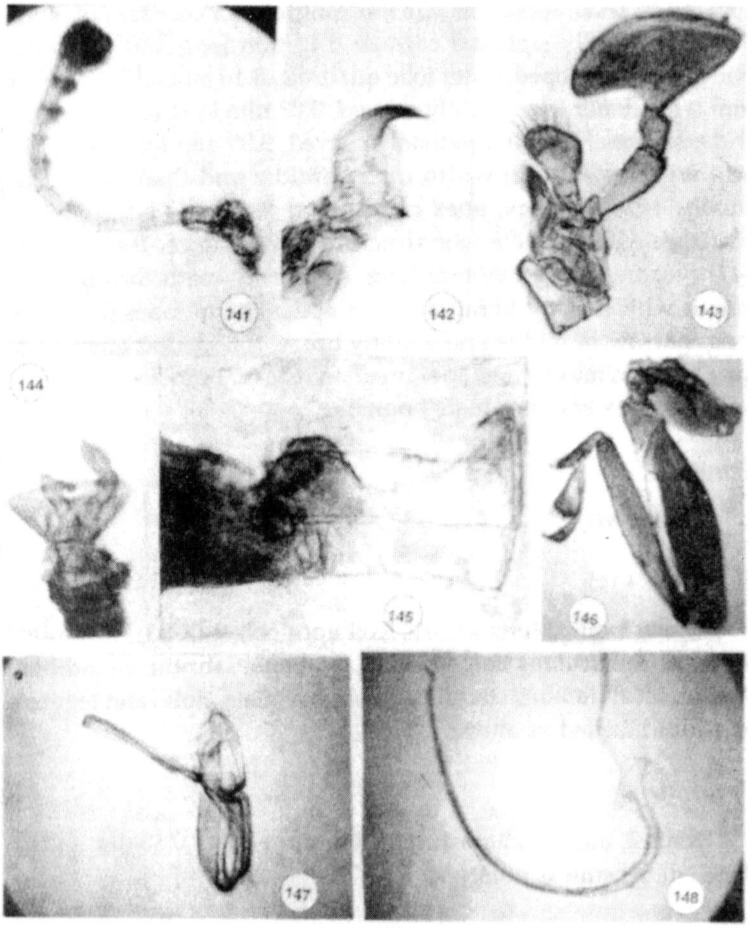

Illeis sathei **sp. nov.**

Fig. 141 : Antenna

Fig. 142 : Mandible

Fig. 143 : Maxilla

Fig. 144 : Labium

Fig. 145 : 1st abdominal sternite

Fig. 146 : Hind leg

Fig. 147 : Tegmen

Fig. 148 : Sipho

above species by having following characters :

i) Elytra densely punctate towards outer margin, on body shows black stripe,

ii) Pronotum without spots,

iii) Head dorsally black and laterally covered by setae.

iv) Mandible with sharply pointed apical bifid tooth and basal tooth abruptly pointed,

v) Scape and terminal segment of antenna with long numerous setae,

vi) Siphonal apex curved with characteristic structure on curvature,

vii) Lateral lobes of tegmen with setae on middle and apex part and

viii) Flageller formula.

Illeis darbarii sp. nov.

(Plae-2 Fig. 20; Plate-20, Figs. 149-157)

Female

Body oval, 6.21 mm long and 5.27 mm wide, dorsum weakly convex; pronotum 1.37 mm long and 2.74 mm wide; antenna 1.94 mm long; elytra 3.37 mm long and 2.23 mm wide, coarsly punctured and without spots; spermatehca 0.35 mm long and 0.10 mm wide, broadly "U" shaped.

Head

1.47 mm long and 1.68 mm wide; subrounded; dorsum pubescent; posterior border coated by hairs; frons black, finely pubescent, fronto clypeal suture indistinct; clypeus slightly broad and expanded infront of compound eyes; surface of eye glabrous, interocular distance 0.63 mm; gular sutures slightly disticnt; tentorium with prallel arms. *Antenna (Fig. 149)* : 11 segmented, 1.94 mm long serrated and thin; scape large, longer than width with 7 long setae; pedicel as long as wide, sub quadrate; 1st segment longer than wide with 2 setae; 2nd and 3rd segments more or less equal; 7th and 8th segments distinctly longer than wide; 9th segment elongated, oval and truncate to apical side and covered by long setae.

Flagellar formula : SL/W=1.24, PL/W=1.6, 1L/W=3.5, 2L/W=2.6, 4L/W=2, 6L/W=1.6, 7L/W=1.6, 8L/W=1.3, 9L/W=1.08.

Labrum : flat, 0.32 mm long and 0.63 mm wide, oval plate like, lateral sides rounded with elevations in the middle and densely covered by setae. *Mandible (Fig. 150)* : somewhat triangular, 0.31 mm long and 0.27 mm wide; apical tooth bifid, reddish, sharply pointed with 0.21 mm depression; basal tooth absent; membranous prostheca 0.29 mm. *Maxilla (Fig. 152)* : cardo as wide as stipe, 0.21 mm long and 0.25 mm wide; stipe more longer than width, 0.31 mm long; galea large, 0.21 mm long, spatulate and covered by thick long setae; lacinia simple, 0.10 mm long, leaf like with long setae; maxillary palpus 4 segmented, 0.99 mm long, basal segment small; 2nd segment longer than 3rd 0.35 mm long; 3rd segment small. 0.14 mm long, quadrate, as wide as long; terminal segment large, 0.33 mm long and 0.84 mm wide, transverse, outer and inner borders expanded, capitulate, with rough hairy surface. *Labium (Fig. 151)* : post mentum fused with gula; mentum squarish, 0.12 mm long and 0.18 mm wide; prementum somewhat tirangular, 0.23 mm long and 0.48 mm wide, apically broad and covered by numerous fine hairs; labial palpus 3 segmented, 0.37 mm long, basal segment small; 2nd and 3rd segments more or less equal, 3rd segment slightly broad at base with narrowly and bluntly pointed apex.

Thorax

1.68 mm long and 3.16 mm wide; pronotum yellowish, 1.37 mm long and 2.74 mm wide with 2 black spots on posterior margin, spots small and subquadrate, lateral corners bluntly rounded, surface very finely punctured and less shegreened; prosternal carinae straight, except posterior end; anterior margin of mesosternum acurately emarginate, metendosternite "Y" shaped with pair of fulcra; scutellum tirangular. *Elytra (Fig. 20)* : pale yellowish, 3.37 mm long and 2.23 mm wide, anterior corners rounded, outer border inclined towards apex and pointed, reticulate and without spots; elytral epipleuron horizontal without fovae and not reaching elytral apex. *Hind leg (Fig. 153)* : coxa stout, 1.16 mm long and 0.42 mm wide, elongated and cylindrical; trochanter 0.42 mm long and 0.35 mm wide, triangular, cap like with denticulate process on inner side; femur 1.79 mm long and 0.65 mm wide, large, broad in middle with coating of setae; tibia 1.89 mm long and 0.29 mm wide, cylindrical with coating of tibial spines on the inner side; tarsi 3 segmented, 2nd segment 0.53 mm long with setae on anterior side; terminal segment 0.78 mm long gradually broad towards the apex and terminate into bifid claw; claws pointed, 0.21 mm long, immediately curved with some what quadrate basal tooth.

Abdomen

2.59 mm long and 2.80 mm wide, yellowish, six sternites visible lateral borders covered by thick hairs; coxal line nearly straight; femoral line on 1st abdominal sternite incomplete, reaching to posterior margin of sternite and area around femoral line smooth (Fig. 154). *Female genitalia* : 9th tergite, a pair of 9th pleurite and a pair of hemisternites together form ovipositor; hemisternite oval, elongate with stylus. *Spermatheca (Fig. 155)* : broadly "V" shaped, 0.35 mm long and 0.10 mm wide, slightly broad at base and narrow towards apex, surface shows markings with depression on lateral side.

Male

5.27 mm long and 4.22 mm wide, similar like female except body size. *Male genitalia* : 9th tergite "U" shaped; 10th tergite transverse and covered by setae; apophysis elongated with rounded apex. *Sipho (Fig. 157)* : 3.26 mm long, siphonal capsule 0.16 mm long and 0.50 mm wide, well developed, outer lobe 0.18 mm long and 0.21 mm wide somewhat quadrate, inner lobe 0.08 mm long and 0.16 mm wide, smaller than outer lobe and slightly curved, notch between lobes less prominant; siphonal tube 3.10 mm long and 0.08 mm wide immediately but widely curved, gradually width decreased and ended bluntly. *Tegmen (Fig. 156)* : lateral roundly ended with coating of thick long setae; median lobe triangular, 1.37 mm long, knife like with slightly hook like apex; basal place squarish; tegmen strut attached to basal piece, broad at apex, 1.05 mm long.

PLATE - 20

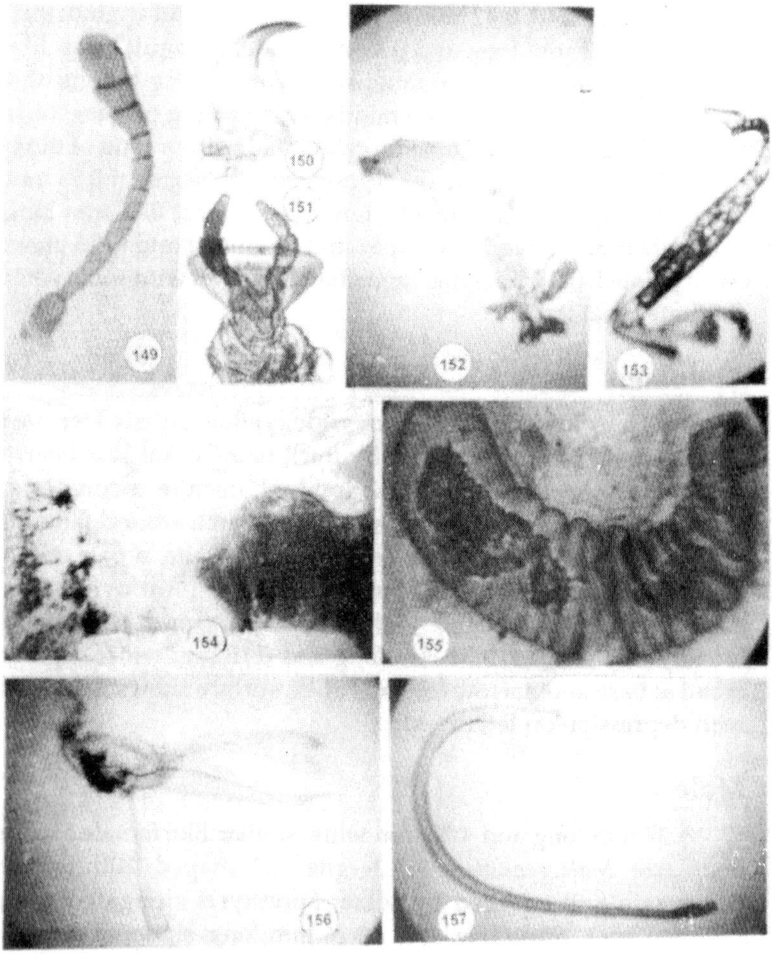

***Illeis darbarii* sp. nov.**

Fig. 149 : Antenna

Fig. 150 : Mandible

Fig. 151 : Labium

Fig. 152 : Maxilla

Fig. 153 : Hind leg

Fig. 154 : 1st abdominal sternite

Fig 155 : Spermatheca

Fig. 156 : Tegmen

Fig. 157 : Sipho

Holotype

Female, India : Maharashtra, Miraj, Coll., 25-VII-93, Y.A. Bhosale, on Salad (*Spinacia oleracea* Linn.), antenna, labrum mandible, labium, maxilla, 1st abdominal sternite, hind leg, spermatheca, last terga of male, sipho and tegmen on slides labelled as above.

Paratype

Female 38, male 12, sex ratio (male : female), 1:3.16, data same as in holotype, coll., from 25-VII-93 to 20-XI-93, Y.A. Bhosale, on Salad (*S. oleracea*) and Jowar (*Sorghum vulgare* Pers.) India : Mahrashtra, Miraj.

Prey

Aphids, *Myzus persicae* Sulzer, *Rhopalasiphum maidis* Fitch.

Prey Plant

Salad, *S. oleracea,* Jowar, *S. vulgare.*

DISCUSSION

This species resembles with *Illeis bistigmosa* by pronotal spots, antenna, labium, 1st abdominal sternite. It also resembles with *I. indica* by having lateral lobes of tegmen more or less equal like median lobe or slightly shorter than median lobe. However, it differs from the above species by having the following characters :

(i) Mandible with bifid and apically sharp pointed tooth,

(ii) Terminal segment of maxillary palpus more transverse,

(iii) Spermatheca with markings with depression on lateral sides,

(iv) Median lobe of tegmen triangular with hooked apex,

(v) Siphonal tube immediately and widely curved with blunt apex.

(vi) Area around femoral line on 1st abdominal sternite smooth and

(vii) Flagellar formula.

Genus *Horniolus* Weise, 1900

Homiolus Weise, 1900, *Dtsch. ent.* Z. 442

Type species : *Horniolus dispar* Weise, 1900.

The genus *Horniolus* was proposed by Weise in 1900, for the inclusion of a Ceylonese species, *H. dispar.* Under this genus following species are described; *H. fortunatus, H. kyashuensis, H. amamensis* and *H. okinawensis* (Miyatake, 1963). Recently Miyatake, 1976 described the five new species under this genus *viz., H. hisamatsni, H. siamensis, H. nigripes* and *H. bimaculatus* and *H. vietnamicus.*

The genus is represented by following characters: body large, oblong, oval and moderately convex on dorsal side; head black with reddish anterior part; pronotum black, sometimes with two discal spots; elytra with two pairs of bands; prosternal carinae inverted "Y" shaped; femoral line on first abdominal sternum complete, very strongly bending; sipho of male genitalia short, distinctly convergent apically median lobe of tegmen boat shaped.

Horniolus mirajensis sp. nov.

(Plate-2, Fig. 18; Plate-21, Figs. 158-165)

Male

3.20 mm long and 2.20 mm wide, dorsum convex; pronotum 1.05 mm long and 2.06 mm wide, black with anterior yellowish brown border; antenna 1.22 mm long; elytra yellowish brown, with large punctures and 2 large black spots.

Head

1.11 mm long and 1.60 mm wide, black, glabrous, completely inserted into pronotum, quadrate with rounded lateral borders; frons black, glabrous with rough surface, fronto clypeal suture indistinct; clypeus slightly projected anteriorly with shiny black rough surface, interocular distance 0.73 mm; gular suture indistinct; tentorium with 2 arms. *Antenna (Fig. 158)* : 11 segmented, clavate, 1.22 mm long,yellowish; scape large, somewhat squarish with 3 long setae; pedicel smaller than scape, as wide as long with one seta; 1st segment slighlty longer than width; 2nd to 6th more or less equal; last three segments form club, 9th segment larger than preceding segments and covered by few long hairs.

Flagellar formula : SL/W = 1.50, PL/W = 1.25, 1L/W ≐ 2.34, 2L/W = 1.67, 7L/W = 1, 8L/W = 0.84, 9L/W = 1.34.

Labrum (Fig. 162) : 0.33 mm long and 0.65 mm wide, elongately oval with rounded lateral borders, posterior border, depressed in middle, antero-lateral side covered by setae. *Mandible (Fig. 161) :* subquadrate, 0.46 mm long and 0.31 mm wide; apical tooth bifid with depression 0.16 mm; basal tooth broad at base, slighlty curved, bluntly ended; membranous prostheca 0.23 mm. *Maxilla (Fig. 159)* : cardo somewhat quadrate, 0.20 mm long and 0.30 mm wide; stipe longer than width; galea large, 0.22 mm large, hood like with setae; lacinia as long as galea, simple withsetae; maxillary palpus 4 segmented; basal segment small. 0.10 mm long; 2nd longer than 3rd, 0.29 mm long; 3rd small, 0.18 mm long with single seta; terminal segment large, 0.41 mm long and 0.37 mm wide, apically truncate, securiform. *Labium (Fig. 160) :* post mentum and gula fused; mentum somewhat triangular, 0.25 mm long and 0.19 mm wide; prementum quadrate, 0.27 mm long and 0.25 wide, apically bordens by hairs; labial palpus 3 segmented. basal segment very small. 0.08 mm; 2nd broader at apex, 0.10 mm; terminal segment small, 0.12 mm elongated and roundly ended.

Thorax

1.15 mm long and 2.10 mm wide; pronotum black, 1.05 mm long and 2.06 mm wide, glabrous, smooth, depressed in middle of anterior side, antero-lateral corners bluntly pointed and not curved, anterior border yellowish brown; prosternum carinate, pair of carinae diverge anteriorly; mesosternum broader than long; scutellum black triangular. *Elytra (Fig. 18) :* 2.01 mm long and 1.15 mm wide, yellowish brown with close punctures, elytral base broader than pronotum, inner border black, suture black; elytron with 2 large black spots; anterior spot broad, larger towards outer margin, narrow in middle and slightly broad towards the inner margin and fuse with its ounter part on another elytron; posterior spot same as anterior spot, apical tip of elytron black; elytral epipleuron transverse and reach to apex. *Hind leg* (Fig. 163) :

Yellowish brown, coxa small and stout, 0.89mm long and 0.79mm wide; trochanter small, quadrate with denticulate process; femur elogated, 1.23mm long and 0.89mm wide, broad in middle; tibia elongated, 1.30mm long and 0.85mm wide with setae on inner side with prominant tibial spur; tarsi 3 segmented; 1st segment elongated, 0.90mm long with broader apex and with setae; 2nd segment triangular 0.98 mm long with few setae; terminal segment elongated, 1.25 mm long gradually broad, towards apex and terminate into bifid claw; claws pointed, 0.40mm very much prominent with basal tooth.

Abdomen

2.02mm long and 2.15mm wide, dorsal side black in middle and yellowish brown on the lateral side, smooth; ventral side hairy; six sternite visible; coxal line on 1st abdominal segment straight and complete; femoral line slightly angulate and reaching to posterior side. *Male genitalia* : 9th terga "U" shaped; 10th terga transverse and covered by setae; apophysis elongated, equal in width throughout length and without broad and bifid base. *Sipho* (Fig. 165) : siphonal capsule well developed, outer lobe quadrate, inner lobe bluntly rounded and slightly curved, notch between lobes less prominent; siphonal tube immediately curved, apex of tube with very peculiar structure, terminate into elongated slightly curved pointed structure. *Tegmen* (Fig. 164) : lateral lobes as long as median lobe, runs parallel, similar in width, apex rounded covered by long fine setae; median lobe 0.89mm long, broad at base, boat shaped, apex narrowed and bluntly pointed.

Female

Unknown.

Holotype

Male, India : Maharashtra, Miraj, Coll., 20-VIII-93, Y.V. Bhosale on sunflower (*Helianthus annus* Linn.), antenna, labrum, maxilla, mandible, sipho and tegmen on slides labelled as above.

PLATE - 21

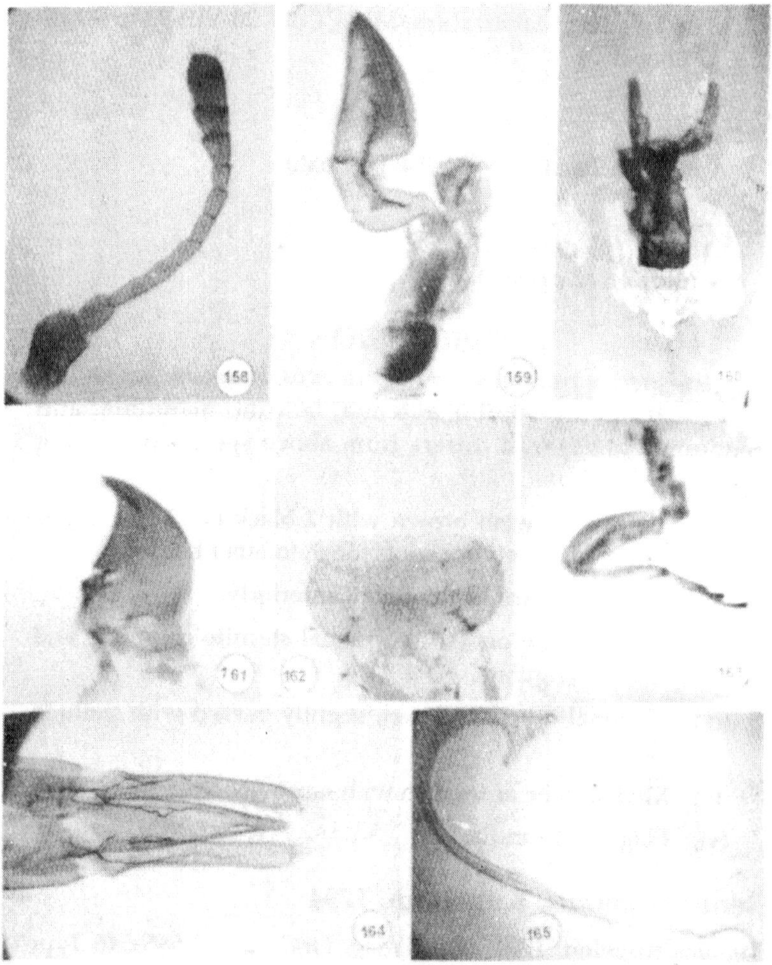

Horniolus mirajensis sp. nov.

Fig. 158 : Antenna
Fig. 159 : Maxilla
Fig. 160 : Labium
Fig. 161 : Mandible
Fig. 162 : Labrum
Fig. 163 : Hind leg
Fig. 164 : Tegmen
Fig. 165 : Sipho

Paratype

Male 4, India : Maharashtra, Miraj, Coll., 20-VIII-93 to 5-XII-93, Y.A. Bhosale.

Prey

Aphid, *Dactynotus compositae* Theobald.

Prey Plant

Sunflower, *H. annus*.

DISCUSSION

This species runs close to *H. fortunatus. H. amamensis* and *H. nigripes* by having similar antenna, labrum, mandible, and pronotum. However, it differs from above species by having following characters :

(i) Elytra yellowish brown with 2 black band like spots, suture black, elytral spots touch to inner border,

(ii) Prosternal carinae divergent anteriorly,

(iii) Femoral line on 1st abdominal sternite complete and slightly angulate,

(iv) Siphonal tube elongated, slightly curved with pointed apex,

(v) Median lobe of tegmen with narrowed apex and

(vi) Flagellar formula.

Genus *Scymnus*, Kugelann, 1794

Scymnus Kugelan, 1794, *Neues Mag/ Liegh. Ent.* 1:545-546 Type species : *Scymnus nigritus* Kugelann, 1794.

The genus *Scymnus* is erected by Kugelann in 1794 and he himself sub divided it into 4 subgenera *viz;* Scymnus, Pullus, Neopullus and Nipponopullus. In all 112 species have been described under this genus.

This genus is characterised by body small, usually short oval, dorsum moderately convex and pubescent; head capsule transversely quadrate with rounded corners; clypeus nearly as broad as or narrower than distance between both antennal insertions; antenna usually 11 segmented, some times 10 segmented; pronotum weakly excavated prosternum normal, "T"-form, usually with distinct carinae; abdomen with six visible segments; legs moderately stout and short, tibia flattened externally and without distinct spur.

Sub Genus *Scymnus* Kugelann, 1794

Under this sub genus 13 species *viz; Scymnus.* (*Scymnus.*) *nebulosus* LeConte, 1852; *S. (S.) paganus* Lewis, 1896; *S. (S.) circumspects* Horn, 1895; *S. (S.) indianensis* Weise, 1929; *S. (S.) nubilus* Mulsant 1931; *S. (S.) miyamotoi* Nakane, 1959; *S. (S.) nigrosuturalis* Kamiya, 1961; *S. (S.) marinus* Kamiya, 1961; *S. (S.) tsuhimaensis* Sasaji, 1970; *S. (S.) paracanus,* Chapin, 1973; *S. (S.) apicanus* Chapin, 1973; *S. (S.)* sp. 1 and sp. 2. Bhaskar, 1992 have been described.

The sub genus *Scymnus* is recognized by the following characters; antenna always 11 segmented; prosternal carinae distinct; femoral line of first abdominal sternum incomplete, its lateral part distinctly recurrent basally, if continued it would reach to the basal margin of the sternum area around femoral line, more or less smooth or more sparsely punctate.

Scymnus (Scymnus) indica sp. nov.

(Plate-2, Fig. 25; Plate-22, Figs. 166-173)

Female

Body small, shortly elongate oval, 2.53 mm long and 1.62 mm wide; dorsum pubescent and moderately convex; pronotum somewhat triangular and not covered the head; antenna 0.42 mm long; elytra punctate, pubescent; elytral suture black.

Head

Somewhat transversely squarish with rounded lateral borders, 0.64 mm long and 0.84 mm wide; frons yellowish, fronto clypeal suture indistinct, clypeus slightly projected or expanded in front of compound eyes; eyes black, laterally projected, interocular distance 0.42 mm; dorsal surface densly hairy; gular sutures run parallel. *Antenna (Fig. 166)* : 11 segmented, clavate, 0.42 mm long; scape large more longer than width, curved at base and with two setae; pedicel as long as broad, somewhat quadrate, 1st segment longer than width; 2nd to 5th more or less equal; 6th and 7th slightly larger than preceeding segments and more or less equal; 8th segment longer than other segments, somewhat quadrate; 9th or terminal segment smaller than 8th, exactly quadrate; 8th and 9th segments covered by fine setae.

Flagellar formula : SL/W=2, PL/W=1, 1L/W=2, 2L/W=1, 6L/W=1, 8L/W=1 9L/W=1.

Labrum (Fig. 169) : small, flat with rounded lateral sides, 0.16 mm long and 0.27 mm wide, covered by long fine setae. *Mandible (Fig. 167)* : small, somewhat quadrate, 0.22 mm long and 0.16 mm wide, apical tooth bifid, depression between bifid apical tooth 0.06 mm; basal tooth slightly curved and bluntly pointed; membranous prostheca 0.10 mm. *Maxilla (Fig. 168)* : cardo as broad as stipe; stipe 0.27 mm long and 0.16 mm wide; galea spatulate, 0.10 mm long; lacinia simple, 0.10 mm long flat, small, elongated and covered by setae; maxillar palpus 4 segmenteed, 0.35 mm long, basal segment small; terminal segment 0.16 mm long and 0.10mm wide roughly quadrate, both sides run parallel with slightly truncate apex. *Labium* : post mentum fused with gula; mentum small, triangular, 0.10 mm long and 0.16 mm wide, anterior corner comes outwardly; prementum somewhat cylindrical, 0.08 mm long and wide with hairs on apical side; labial palpus 3 segmented, 0.16 mm long, basal segment small, terminal segment longer than 2nd segment, broad in middle and bluntly pointed.

Thorax

1.26 mm long and 1.58 mm wide; pronotum somewhat triangular, smooth with a black distinct marking at the middle of the basal area, the black marking covering the sutural area of elytra, slightly broadened in middle and extending to the apex, lateral corner subangulate; prosternal carinae straight and slightly diverging posteriorly. *Elytra (Fig. 25)* : 1.89 mm long and 1.26 mm wide yellowish red, granulate, sparsely but strongly punctate with hairs, inner border black, base straight, apically rounded; ventral side with 2 zig zag lines, one in middle and other towards the inner border; elytral epipleuron narrow, horizontal and reaching to apex. *Hind leg (Fig. 171)* : coxa stout, elongated; trochanter small subquadrae with slight denticulate process; femur broad, 0.63 mm long and 0.21 mm wide, elongated, similar in width; tibia 0.63 mm long and 0.08 mm wide, cylindrical with setae on lateral side; tarsi cryptotetramerous, 1st segment 0.16 mm long conical, narrow at base; 2nd 0.14 long elongated; 3rd 0.04 mm long, quadrate; terminal segment 0.12 long gradually broad towards the apex and terminate into bifid claw; claw pointed and 0.06 mm long.

Abdomen

1.05 mm long and 1.37 mm wide, dorsum and ventrum black; ventrum pubescent; lateral borders densly covered by setae; six sternites visible; coxal line on 1st abdominal sternite complete, femoral line incomplete and sub-angulate, area around femora long, densely and coarsely punctate (Fig. 170). *Female genitalia* : hemisternites elongated, slightly broad at base, bluntly narrowly pointed with stylus. *Receptaculum seminis (Fig. 173)* : small, broadly "V" shaped; cornu bluntly rounded, slightly broad at base, base with close markings, middle part with large markings; inner surface "V" shaped; cornu 0.12 mm long, base 0.16 mm long.

Male

Same as female expect body size, small, 2.21 mm long and 1.39 mm wide. *Male genitalia* : 9th tergite slightly "U" shaped; 10th tergite transverse; 9th pleurite attached to 9th tergite; 9th sternite small; apophysis with swollen part at apex. *Sipho* : slender, basal part halfly rounded, slightly swollen in middle, siphonal capsule with long inner process and small outer process. *Tegmen (Fig. 172)* : stout; lateral lobe boat like with pointed and slightly longer than median lobe with setae on apex, 0.23 mm long; median lobe 0.20 mm long tegmen stout elongated with slightly broad apex, 0.54 mm long.

Holotype

Female, India : Maharashtra, Miraj, coll., 25-I-94, Y.A. Bhosale, on Sunflower, (*Helianthus annus* Linn.), antenna, labrum, mandible, maxilla, hind leg, 1st abdominal sternite, spermatheca, sipho, tegment and last abdominal terga of male on slides labelled as above.

Paratype

Female 5, male 2, sex ratio (male: female), 1:2.5, data same as in holotype coll., from 23-I-94 to 5_III-94 Y.A. Bosale, India : Maharashtra, Miraj.

Prey

Aphid, *Dactynotus compositae* Theobald

Prey Plant

Sunflower, *H. annus.*

DISCUSSION

This species runs close to *Scymnus (Scymnus) nigrosuturalis* by having antenna, maxilla, labrum, pronotal basal protion and elytral sutural margin *S. (S.) indica sp. nov,* also resembles with S. (S.) sp. 1 by last terga of male, tegmen and hemisternites of ovipositor. It also resembles with *S. (S.) indianensis* by having tegmen, sipho with out fish hooking projection. But it differs from *S. (S.) nigrosuturalis* by having following characters :

- i) Elytra on its ventral side shows two zigzag markings,
- ii) Area around the femoral line on 1st abdominal sternite is densely and closely punctate,
- iii) Prosternal carinae straightly divergent posteriorly,
- iv) Spermatheca with close and large markings on surface,
- v) Median lobe of tegmen terminate into slightly pointly hook like structure,
- vi) Apex of apophysis largely swollen and
- vii) Flagellar formula and from *S. (S.)* sp. 1 by :
    - i) Pronotum basally shows black margin which continued on elytral suture upto apex,
    - ii) Area around femoral line densely and closely

PLATE - 22

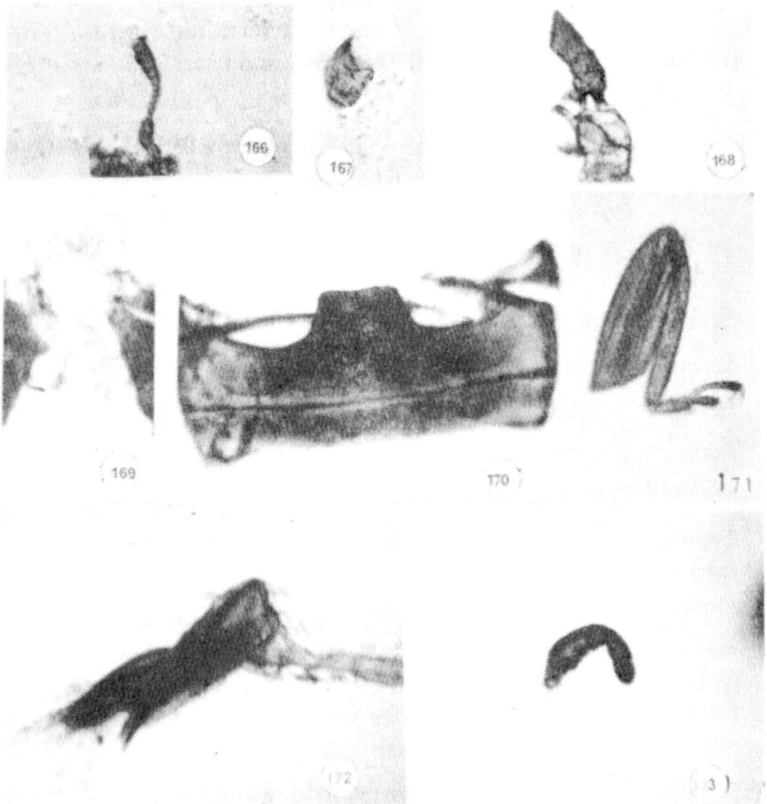

Scymnus (scymnus) indica **sp. nov.**

Fig. 166 : Antenna
Fig. 167 : Mandible
Fig. 168 : Maxilla
Fig. 169 : Labrum
Fig. 170 : 1st abdominal sternite
Fig. 171 : Hind leg
Fig. 172 : Tegmen
Fig. 173 : Spermatheca

punctate,

iii) Prosternal carina straightly divergent basally and

iv) Median lobe of tegment terminate into alightly pointed hook like structure and from *S. (S.) indianensis* by :

(a) Pronotum black at base and black portion continued on elytral suture,

(b) Apophysis with swollen part at apex and

(c) Elytra yellowish red.

Sub Genus *Pullus* Mulsant, 1846

Scymnus (Pullus) Mulsant, 1846.

Hist.Nat.Colept.Fr. Securing : 241.

Pullus; Motschulsky, 1866.

Bull.Soc. Nat. Moscou. 39 : 426.

Type Species : *Coccinella fasciata* Fourcroy, 1785 = *Scymnus (Pullus) subvillosus* (Goeze/1777).

The *Pullus* is the largest subgenus of the genus *Scymnus*. Casey reviewed the whole group in 1899. Wingo (1952) studied 24 species from upper Mississippi basin and described 8 new species. Sasaji (1968, 71) described 29 species from Japan and Chapin (1974) described 12 species from Louisiana.

From India 16 species have been described by various authors, Mohan Rao (1958) reported 3 species from Northern Circars and very recently Bhaskar (1992) described 6 species from Karnataka. Other species described from India includes *S. (P.) bourdilloni, S. (P.) testacecollis* and *S. (P.) hingstoni* from third mount everest (Kapur, 1963), *S. (P.) brunnescens* Motschulsky and *S. (P.) nymphaeus* from Calcutta (Kapur and Munshi, 1970) and *S. (P.) quadrillum* Motschulsky from Goa (Kapur, 1972).

The subgenus *Pullus* is separable from the subgenus *Scymnus* by the complete femoral line which is actually reaches to the basal margin of the sternum; prosternal lobe with 2 carinae, fifth sternum of male usually emarginate, male genitalia symmetrical or asymmetrical.

Scymnus (Pullus) marathi sp. nov.

(Plate-2, Fig. 27; Plate-23, Figs. 174-182)

Female

Body small, reddish brown, finely pubescent, oval, strongly convex on dorsal side, 2.47 mm long and 2.0 mm wide , pronotum reddish brown, finely pubescent, 0.84 mm long and 1.30 mm wide; antenna 0.33 mm long; elytra 2.21 mm long and 1.37 mm wide, granulate, reddish brown; spermatheca 0.37 mm long, without markings on surface.

Head

Somewhat rounded, 0.63 mm long and 0.84 mm wide; frons reddish, glabrous, fronto clypeal sutre indistinct; clypeus anteriorly slightly projected; lateral borders covered by hairs; compound eyes very much prominant, black, shiny, inter ocular distance 0.35 mm; gular sutures indistinct. *Antenna (Fig. 174)* : 11 segemented, clavate 0.33 mm long; scape large, broad, bent at base with three long setae; pedicel as wide as scape with two setae; 1st segement longer than width; 2nd to 5th similar and smaller than first segment; 6th segment more wider than length, first segment; transverse and smaller than other segments; 7th segment broad; 8th segment larger than all other segments; 9th segment partly embeded in 8th and remain cap like; six to nine segments covered by long and fine setae, more setae on 9th segment.

Flagellar formula : SL/W=1, PL/W=1, 1L/W=2, 2L/W=0.06, 6L/W=0.5, 7L/W=0.33, 8L/W=0.66, 9L/W=0.5.

Labrum : small, plate like 0.14 mm long and 0.25 mm wide, covered by setae on anterior and lateral side. *Mandible (Fig. 175)* ; somewhat triangular, 0.23 mm long and 0.25 mm wide at middle; apical tooth slightly curved and pointed; membranous prostheca 0.08 mm; outer border concave at base and then strongly convex. *Maxilla (Fig. 176)* : cardo 0.09 mm long and 0.08 mm wide, as broad as stipe : stipe 0.12 mm long and 0.08 mm wide : lacinia 0.06 mm long, simple with setae; maxillary palpus 4 segmented, 0.27 mm long, basal segment very small; 2nd large, more elongated than

3rd; 3rd as broad as long; terminal segment large, elongated, obliquely apically truncate and securiform. *Labium (Fig. 177) :* submentum fused with gula; mentum large, 0.10 mm long and 0.21 mm wide, lateral sides anteriorly projected outwardly; prementum 0.12 and long and 0.08 mm wide, elongated and anteriorly covered by setae; labial palpus 3 segmented.

Thorax

1.68 mm long and 1.89 mm wide; pronotum reddish brown, finely pubesent, sparsely puncate, 0.84 mm long and 1.30 mm broad, anterior border broadly excavated, antero lateral corners angulate, posterior border slightly concave, lateral and posterior border covered by setae; prosternum with parallel carine covergent anteriorly; mesosternum and metasternum black; scutellum reddish brown and triangular. *Elytra (Fig. 27) :* 2.21 mm long and 1.37 mm wide, brown, finely pubescent with backwardly sloping hair (setae), inner border black, basal border straight, apical side bluntly rounded; elytral epipleuron narrow and horizontal. *Hind leg (Fig. 179) :* coxa 0.37 mm long and 0.18 mm wide, elongated and stout; trochanter 0.18 mm long and wide, denticulate processa nd onner side; femur 0.54 mm long and 0.25 mm wide broad in middle and covered by setae; tibia elongated 0.54 mm long and 0.12 mm wide more towards apical region; tarsi tetramerous, 0.33 mm long 1st segment triangular, 0.16 mm long, 2nd 0.14 mm long elongated, 3rd very small, 1.04 mm long; terminal segment 0.12 mm long, elongated and terminate into bifid claw; claw pointed, 0.12 mm long; curved and without basal tooth.

Abdomen

1.26 mm long and 1.89 mm wide; ventrum reddish brown; six sternites visible, lateral margins covered by setae, coxal line on 1st abdominal sternite complete, femoral line curved and anteriorly disappears, area around the femoral line finely punctate (Fig. 178); 1st abdominal sternum in male modified in the middle. *Female genitalia :* hemisternites elongated, triangular, bluntly pointed and with a stylus. *Receptaculum seminis (Fig. 182) :* small, broadly "U" shaped, cornu and nodulus broad with dark brown tinge and close markings, middle protion yellowish and without markings, nodulus base slightly broad than cornu, cornu 0.14 mm long and 0.08 mm broad, nodulus 0.14 mm long and 0.06 mm wide, total length of spermatheca 0.37 mm.

Male

Same as female except body size, 2.15 mm long and 1.89 mm broad. *Male genitalia :* 9th tergite "U" shaped; 10th tergite slightly transverse; apophysis elongated slightly broad at apex. *Sipho :* stout, basal portion curved semicurcularly, siphonal capsule well developed, inner lobe distinctly longer than outer lobe, apex with inner triangular extension and with long thread like process at its apex. *Tagmen (Figs. 190, 191) :* stout; lateral lobes longer than median lobe, lateral lobes and apically covered by long dense setae, 0.31 mm long; median lobe small basally broad, apical portion pointer and hook, like 0.12 mm long and 0.10 mm wide; basal piece quadrate; tegmen stout elongated, more longer than other parts, 0.54 mm long.

Holotype

Female, India : Maharashtra, Kolhapur, Coll., 12-II-94, T.V. Sathe, on Sunflower (*Helianthus annus* Linn.), antenna, labrum, mandible, maxilla, hind leg, 1st abdminal sternite, spermatheca, sipho and tegmen of male on slides labelled as above.

Paratype

Female 20, male 9, sex ratio (male:female), 1:2.22, data same as in holotype, coll. from 12-II-94 to 20-III-94, Y.A. Bhosale, on Jowar (*Sorghum vulgare* Pers.), India : Maharashtra, Kolhapur and Miraj.

Prey

Aphids, *Dactynotus compositae* Theobald, *Rhopalsiphum maidis* (Fitch).

Prey Plant

Sun flower *H. annus.* Jowar *S. Vulgare.*

PLATE - 23

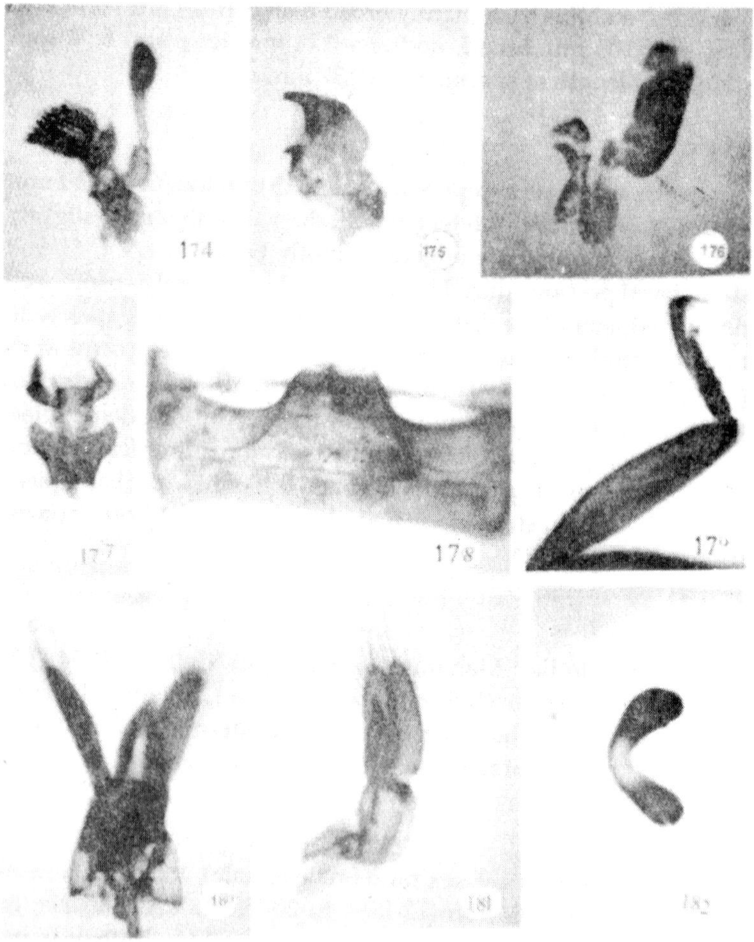

Scymnus (Pullus) marathi **sp. nov.**

Fig. 174 : Antenna
Fig. 175 : Mandible
Fig. 176 : Maxilla
Fig. 177 : Labium
Fig. 178 : 1st abdominal sternite
Fig. 179 : Hind leg
Fig. 180 : Tegmen
Fig. 181 : Tegmen (lateral view)
Fig. 182 : Spermatheca

DISCUSSION

This species resembles to *Scymnus (Pullus) hilaris* Molschulsky by having antenna, labrum, maxilla, prostennal carinae and sipho. However, it difers from the above species by having the following characters :

i) Elytra reddish brown with backwardly sloping hairs, elytra coarsly punctate,

ii) Scutellum reddish brown,

iii) Area around the femoral line on 1st abdominal sternite finely punctate, femoral line disappears anteriorly,

iv) Spermatheca with close markings on basal and apical part and without markings in the middle,

v) Median lobe of tegmen broad in middle and terminate into pointed hook like structure and,

vi) Flagellar formula.

PLATE - 24

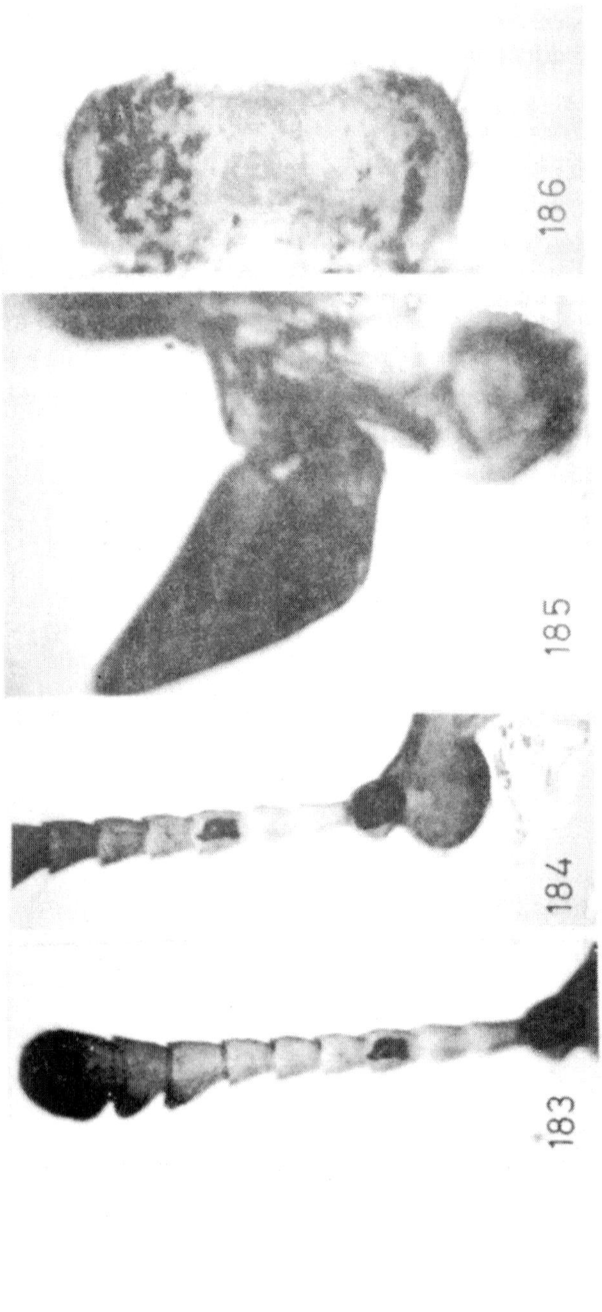

Coccinella yoginii sp. nov.

Fig. 183 : Antenna; Fig. 184 : Antenna (Scape Region);
Fig. 185 : Maxilla; Fig. 186 : Labrum

4

Biology

INTRODUCTION

During last 40 years considerable attention has been paid on classical approaches in the development of biological control techniques, which involves introduction of exotic predators and parasitoids into area in order to supress the pest species population. True identity of the species, biological peculiarities, provision of alternative hosts, nutrition, potential cultural media, manipulation of behaviours, predators, etc. are important steps in biological control. In biological control two distinct categories for agumentation of predators recognised are the inoculative and inductive release (DeRach and Hagen, 1964). There are numerous reviews available on inductive releases in biological control and integreated pest management (DeBach, 1964; Ridgway *et al.*, 1974; Rabb *et al.*, 1976; etc.).

A very large number of predaceous species occur throughout the world possessing fascinating biologies and frequently function as pest population densities determinates. Hence they have tremendous economic importance and therefore, largely used in biological control of pests.

The family coccinellidae is figured as one of the prominant group of predators. The majority of coccinellids are of importance as predators on aphids, scales, mealy bugs, jassids, spiders, mites and the eggs and larvae of other insects. However, the subfamily epilachninae is phytophagous.

The structure and number of larval instars have great significance in identification of the species. General criteria for taxonomic assessment is adult characteristic. However, the reliable understanding of the species would be more correct by studying their immature forms. In all individuals there occur changes in the patterns of characters during embryonic development.Such changes are quite obvious in groups where the individual passe quite different morphological stages. The taxonomic description is based not only on the morphological characters of the adult but also on the sum total of the characters of all stages. There are various animal groups where the classification is usefully aided by the characters of their immature stages. The biology and biometry of immature foms help in correlating the probable phylogeny of the groups. In past, Boving (1917), Gage (1920), Strouhal (1926), Binaghi (1941), Van Emben (1949), Kapur (1950), Savoiskaya (1955, 1957), Kenneth and Hagen (1962), Kamiya (1965), Witter (1969), Klausnitzer (1969), Storch (1970), Witsack (1971), Hodek (1973), Mangutova (1975), Sinnathamby (1977), Singh and Malhotra (1979), Rao *et al.* (1981), Richards (1981), Anderson (1981), Umeh Ethel-Doris (1982), Radwan and Love (1983), Ghosh (1986), Diraviam and Viraktamath (1991, 1992), Patro and Behera (1992), Rawat *et al.* (1992), etc. have been worked on biology/ biometry of immature forms of coccinellids. The review of literature shows that very little attention is paid on the above aspects of Indian species. Hence present work was carried out with *Illeis darbarii* sp. nov., a predator of aphid, *Rhopadosiphum maidis* (Fitc) on jowar *Sorghum vulgar* Pers. The predator species *I. darbarii* is described and reported for the first time from Maharashta, India.

Material and Methods

The culture of predators and preys were maintained under laboratory conditions (25°C ± 1°C, 55 to 60% R.H. and 12 hr photoperiod) as per the procedure given in the chapter 2.

Laboratory reared (F1 generation) pairs (1 male and 1 female) were confined either in petridish or rearing unit I for their mating, after the isolation period of 10 days, mating easily occurred at the day time. The mated females start egg laying soon after mating. Sufficient number of eggs were then collected and kept in petridishes for observing further development. Sufficient number of 1st, 2nd, 3rd, 4th instars and pupae were collected and considered for their morphological observations and biometrical measurements. Experiments were designed with 20 individuals. Mandibles of 4 instars were collected on the slide, for this, heads of larvae were cut and kept in hot solution of 10 percent KOH for 2 to 5 min and madibles separated. Later the length and width of the mandibles were measured. Head capsules and diameter of spiracles of the larvae were also measured. The results obtained are tested with the help of regression analysis to find out its significance.

All the measurements are taken by caliberated ocular micrometer in a compound microscope and photographs of the immature forms were taken with the help of micro close up lenses.

RESULTS

Egg (Plate- 25, Fig. 187) : The eggs are elongated oval, yellowish and laid in clusters. The eggs hatch out within 4 days after oviposition. The eggs averaged 1.39 mm in length (range 1.0 to 1.43 mm) and 0.54 mm in width (range 0.42 to 0.60 mm).

Instars : There are four instars, all are vesiculate type. Their measurements are recorded in Table-8.

First instar (Plate-25, Fig. 188) : The larvae were emerged through circular opening from eggs. After hatching from the eggs, the larvae stay on the egg shell for one day and eat some eggs, then they crawl about in search of their normal food. The larval body is yellowsih and distinguished into three regions head, thorax and abdomen with anal vesicle. The head is small and rounded, lateral borders greatly rounded, dorsum smooth, posterior border with setae and fronto clypeal suture indistinct. Antenna is 3 segmented last segment hardly noticeable. The mandible is elongately triangular, small with bifid apical tooth. All legs are 3 segmented, tibiotarsus long, wide and covered by numerous spines and terminate into claw. The claw is single, curved and pointed with small basal tooth. The abdomen is long, 8 segmented, gradually narrowed towards posteriorly, each segment is with 4 brownish strumae and lateral elevated portions with 3 to 4 chalazae. The structures of the stumae are weakly developed. The anal vesicle is wide with notch in centre. The first instar lasts for 5 days.

The biometrical studies were made with 20 individuals. The larval body averaged 3.50 mm in length (range 2.98 to 3.85 mm) and 0.68 in width (0.6 to 0.80 mm). The head averaged 0.31 mm in length (0.28 to 0.36 mm) and 0.42 mm in width (0.38 to 0.45 mm). The mandibles averaged 0.12 mm in length (0.09 to 0.16 mm) and 0.08 mm in width (0.06 to 0.10 mm). The thorax averaged 1.26 mm in length (1.16 to 1.36 mm) and 0.78 mm in width (0.6 to 0.80 mm). The prothorax averaged 0.42 mm in length (0.39 to 0.48 mm) and 0.61 mm in width (0.58 to 0.68 mm); mesothorax and metathorax averaged 0.42 mm (0.39 to 0.48 mm) and 0.78 mm in width (0.6 to 0.80 mm) respectively. The strumae on mesothoracic segment averaged 0.12 mm in length (0.09 to 0.16 mm) and 0.21 mm in width (0.19 to 0.26 mm). The mesothoracic spiracles averaged 0.04 mm in diameter (0.02 to 0.06 mm). The coxa a hind leg averaged 0.14 mm in length (0.10 to 0.18 mm) and 0.12 mm in width (0.09 to 0.16), femur averaged 0.44 mm in length (0.39 to 0.48 mm) and 0.10 mm in width (0.08 to 0.15 mm), tibiotarsus 0.61 mm in length (0.56 to 0.63 mm) and 0.06 mm in width (0.02 to 0.08 mm). The claw averaged 0.06 mm in length (0.03 to 0.07 mm). The abdomen was averaged 1.68 mm in length (1.60 to 1.69 mm), the width of 1st four abdominal segments was 0.65 mm (0.58 to 0.68 mm), the width of 5th and 6th segment 0.56 mm (0.50 to 0.58 mm), the width of 7th segment 0.44 (0.39 to 0.46 mm) and that of the last segment was 0.40 mm (0.38 to 0.43 mm). the strumae on 2nd abdominal segment averaged 0.08 mm in length (0.02 to 0.10 mm) and 0.10 mm in width (0.09 to 0.12 mm). The vesicle averaged 0.42 mm in length (0.38 to 0.45 mm) and 0.31 mm in width (0.29 to 0.34 mm).

Second instar (Plate-25, Fig. 189) : The second instar is yellowish, elongated broad in middle and gradually narrowed towards posterior side with 3 distinct regions *viz.*, head, thorax and abodmen with vesicle. The head is rounded, dorsum smooth, lateral borders covered by setae and frontoclypeal suture indistinct. The antenna is 3 segmented with very small last segment. The mandible is elongately triangular. Each thoracic segment is with brownish strumae and lateral corners are with 3 to 4 chalazae with low bases. The strumae are weakly developed. All legs are 3 segmented, tibiotarsus terminate into single claw with basal tooth. The abodmen is 8 segmented, each segment with 4 strumae and lateral elevations with 3 to 4 chalazae. The anal vesicle is with 4 lobed structure with prominant depressions in between lobes. The second instars lasts for 4 days.

The body of larva averaged 5.06 mm in length (4.92 to 5.20 mm) and 1.26 mm in width (0.98 to 1.28 mm). The head averaged 0.42 mm in length (0.39 to 0.46 mm) and 0.54 mm in width (0.50 to 0.56 mm). The mandibles averaged 0.14 mm in length (0.12 to 0.16 mm) and 0.12 mm in width (0.10 to 0.14 mm). The length of thorax averaged 1.89 mm (1.0 to 1.93 mm) and 1.26 mm in width (0.98 to 1.28 mm). Prothorax averaged 0.62 mm in length (0.58 to 0.65 mm) and 1.16 mm in width (0.98 to 1.19 mm), mesothorax 0.62 mm in length (0.58 to 0.66 mm) and 1.16 mm in width (0.98 mm to 1.19 mm) in width (0.98 to 1.19 mm). Metathorax averaged 0.62 mm in length (0.58 to 0.66 mm) and 1.26 mm in width (0.98 to 1.28 mm). The strumae on mesothoracic segment averaged 0.21 mm in length (0.19 to 0.33 mm) an 0.42 mm in width (0.39 to 0.46 mm). The averaged diameter of mesothoracic spiracle was 0.06 mm (0.02 to 0.07 mm). The coxa of hind leg averaged 0.12 mm in length (0.10 to 0.14 mm) and 0.18 mm in width (0.09 to 0.2 mm), femur averaged 0.73 mm in length (0.68 to 0.75 mm) and 0.14 mm in width (0.12 to 0.15 mm) while tibiotarsus 0.73 mm in length (0.70 to 0.75 mm) and 0.10 mm inwidth (0.08 to 0.12 mm). The claw averaged 0.08 mm in length (0.03 to 0.08 mm). The average length of abdomen was 2.63 mm (2.50 to 2.68 mm). The width of 1st segment was 0.94 mm (0.90 to 0.96 mm) and that of last segment was 0.73 mm (0.70 to 0.76 mm). The abdominal strumae and 2nd segment averaged 0.14 mm (0.12 to 0.16 mm) in length and 0.12 mm in width (0.10 to 0.13 mm). The vesicle averaged 0.25 mm in length (0.20 to 0.28 mm) and 0.52 mm in width (0.48 to 0.55 mm).

Third instar : The body of larva is elongated, gradually narrowed towards posterior side, yellowish with 3 prominant regions, head, thorax and abdomen. The head is similar in shape as in last instar. The dorsum is without hairs but lateral borders with very few hairs. The antennal segmentation was also similar to previous one. The mandible is elongately triangular, small with apical bifid tooth. Each thoracic segment is with brownish strumae and their lateral corners are with 3 to 4 chalazae. All legs were with 3 segments, tibiotarsus with single prominant curved and pointed claw. The claw is with prominent basal tooth. The abdomen is 8 segmented, each with well developed 4 strumae while their lateral corners of segments with 2 to 3 chalazae. The vescile of 3rd instar is with 5 lobed structure and with prominent depressions. This stage lasts for 3 days.

The average body length of 20 individuals was 5.69 mm (5.1 to 5.72 mm) and 1.26 mm in width (1 to 1.29 mm). The head averaged 0.73 mm in length (0.68 to 0.74 mm) and 0.84 mm in width (0.79 to 0.85 mm). The mandibles averaged 0.16 mm (0.12 to 0.17 mm) in length and width. The averaged length of thorax was 2.00 mm (1.92 to 2.05 mm) and 1.26 mm in width (1.18 to 1.28 mm). The prothorax averaged 0.69 mm (0.65 to 0.70 mm) in length and 0.94 mm (0.90 to 1.07 mm) in width, mesothorax 0.65 mm in length (0.62 to 0.66 mm) and 1.05 mm in width (1.0 to 1.07 mm) and metathorax 0.65 mm in length (0.62 to 0.66 mm) and 1.26 mm in width (1.18 to 1.28 mm). The strumae on mesothoracic segment averaged 0.27 mm in length (0.22 to 0.29 mm) and 0.48 mm (0.42 to 0.50 mm) in width. The averaged diameter of mesothoracic spiracle was 0.08 mm (0.04 to 0.09 mm). The coxa of hind leg is 0.42 mm in length (0.40 to 0.45 mm) and 0.25 mm in width (0.22 to 0.26 mm), femur 0.97 mm long (0.92 to 0.99 mm) and 0.25 mm in width (0.22 to 0.26 mm) and tibiotarsus 1.01 mm (0.98 to 1.04 mm) in length and 0.16 mm (0.14 to 0.17 mm) in width. The claw length averaged 0.08 mm (0.04 to 0.09 mm). The length of abdomen averged 2.13 mm (2.0 to 2.15 mm), width of 1st five abdominal segments was 1.05 mm (0.98 to 1.07 mm), 6th segment 0.90 mm (0.88 to 0.94 mm) and that of last segment was 0.73 mm (0.70 to 0.75 mm). The vesicle averaged 0.52 mm in length (0.49 to 0.54 mm) and 0.84 mm in width (0.80 to 0.86 mm).

Fourth instar (Plate-25, Fig. 190) : The larval body is elongate, yellowish, gradually narrowed towards posterior side and shows 3 distinct regions head, thorax and abdomen. The vesicle is present on last abdomenal segment. The head of the larva is rounded, dorsum hairy, lateral borders covered by setae, clypeus extended infront with indistinct frontoclypeal suture. The antenna is 3 segmented. The mandible is similar to previous instar except larger in size. Each thoracic segment is with brownish strumae and the borders of the strumae are covered by 2 to 3 chalazae. The larval legs are with 3 segments and tibiotarsus with more spines towards apical side and terminate into prominent claw with quardrate basal tooth. The thoracic region is clearly arched in middle. The abdomenal segmentation is similar to that 3rd instar. Each abdomenal segment with well developed 4 strumae and 2 to 3 chalazae from lateral sides. The vesicle is with 5 lobed structure and with prominant depressions. The stage lasts for 2 days. The fourth instar before ecdysis, remains immobile, does not feed and attached itself to leaves by anal vesicle and undergoes pupation.

The average length and width of larval body was 8.00 mm (6.98 to 8.30 mm) and 3.27 mm (2.11 to 3.29 mm) respectively. The average length of head was 1.05 mm (0.98 to 1.07 mm) and width 1.09 mm (1.0 to 1.1 mm). The length and width of mandibles averaged 0.18 mm (0.16 to 0.18 mm). The average length of thorax was 2.78 mm (2.29 to 2.80 mm) and 3.27 mm in width (3.19 to 3.29 mm). Prothorax averaged 1.0 mm (0.9 to 1.1 mm) in length and 2.11 mm (2.0 to 2.19 mm) in width. Mesothorax was 0.89 mm in length (0.8 to 0.9 mm) and 2.74 mm in width (2.69 to 2.76 mm) and metathorax 0.89 mm in length (0.85 to 0.90 mm) and 3.27 mm in width (3.19 to 3.29 mm). The averaged length of mesothoracic strumae was 0.69 mm (0.60 to 0.70 mm) and 0.67 mm width (0.61 to 0.69 mm). The diameter of the thoracic spiracle was 0.10 mm (0.09 to 0.13 mm). The coxa of hind leg was 0.63 mm in length (0.60 to 0.65 mm) and 0.42 mm in width (0.4 to 0.45 mm), femur, 1.37 mm in length (1.29 to 1.39 mm) and 0.31 mm in width (0.29 to 0.33 mm) and tibiotarsus 1.39 mm in length (1.29 to 1.40 mm) and 0.32 mm in width (0.30 to 0.34 mm). The length of claw averaged 0.12 mm (0.10 to 0.14 mm). The abodmen avraged 4.27 mm in length (4.00 to 4.30 mm), the width of 1st abdominal segment, 4th segment, 7th segment and last segment was 3.16 mm (2.99 to 3.17 mm), 2.95 mm (2.67 to 3.0 mm), 1.89 mm (1.68 to 1.9 mm) and 1.47 mm (1.39 to 1.49 mm) respectively. The strumae on 2nd abdominal segment averaged 0.33 mm in length (0.29 to 0.35 mm) and 0.42 mm in width (0.39 to 0.43 mm). The vesicle averaged 0.94 mm in length (0.89 to 0.98 mm) and 1.86 mm in width (1.19 to 2.29 mm).

Pupa (Plate-25, Fig. 191) : The pupa is yellowish and small. The dorsum is strongly convex. The colour of the pupa does not changed during its development. The pronotum is with 2 black spots. The abdominal region of the pupa is yellowish and without spots. The sexual dimorphism is observed during pupal stage, those in which last abdominal segment is depressed in middle develops into female while those with last abdominal segment straight develops into male. The pupal stage lasts for 4 days. The pupal skeleton remains attached to leaf after the emergence of adults.

The averaged length and width of pupa was 4.85 mm (4.2 to 4.90 mm) and 3.37 mm (31. to 3.40 mm) respectively. The thorax averaged 1.00 mm in length (0.09 to 1.2 mm) and 2.00 mm in width (1.9 to 2.2 mm). The abdomen averaged 2.00 mm in length (1.98 to 2.02 mm) and 3.31 mm in width (3.29 to 3.34 mm).

Adult (Plate -25, Fig. 192) : Body small, oval, 6.21 mm in length (6.0 to 6.29 mm) and 4.27 mm in width (4.0 to 4.29 mm), pale yellow, dorsum weakly convex, ventrum yellow. Head sub rounded, dorsum pubescence, posterior borders covered by hairs, 1.47 mm in length (1.2 to 1.49 mm) and 1.18 mm in width (1.1 to 1.19 mm). Antenna 11 segmented, serrated, 1.94 mm long, scape large with 7 long setae. Mandible with bifid apical tooth. Pronotum yellowish with 2 black spots on posterior margin, lateral corners bluntly rounded, surface very finely punctured and less shagreened; prosternal carinae straight. Elytra pale yellowish with rounded anterior conrners, outer border inclined toward apex and pointed and without spots. Tarsi 3 segmented and terminate into bifid claw with basal quadrate tooth. Coxal line on 1st abdominal sternite complete; femoral line incomplete with smooth area around the line. Spermatheca broadly "U" shaped with markings and depressions on lateral side.

In male, siphonal capsule well developed with outer and inner lobes, siphonal tube immediately curved; lateral lobes of tegmen more or less equal as median lobe, median lobe triangular, knife like with hook like apex.

DISCUSSION

Persual of literature showed that very little is known about the aspects studied here. The subject was first tackled by Boving (1970), and soon afterwards Gage (1920) described 40 species of

American coccinellids with respect to immature stages. Strouhal in 1926 published a paper which was devoted to larvae of tribe coccinellini and psylloborini. Later Binaghi (1941) studied larvae of chilocorini and some species of scymnini. In 1950 Kapur studied larvae of epilachninae. In 1949 Van Emden extended a key for the larvae of 36 species of British coccinellids. The work by Kamiya (1965) was specially interesting in that, on the basis of larvae of Japanese Coccinellidae, the tribal phylogeny of the family was discussed. Klausnitzer (1969) described the larvae of *Lithophilus cannatus* family Lithophilinae, the same author has produced a monograph with a key to 59 species on the Mid European coccinellid larvae. Savoiskaya (1955, 1957) has published a series of work devoted to the morphology and taxanomy (with keys) for 81 species of coccinellid larvae in the Soviet Union.

Singh and Malhotra (1979) studied the biology of *Coccinella undecimpunctata memteriesi* Mulsant, a predator of mustard aphid wherein they observed that the female deposited cigar shaped shiney yellow eggs in clusters mostly on the under surface of leaves and egg hatching was within 2 days, they further reported four instars in the species, the 1st and 2nd instars were with one day duration and pupal period was 3.3 days, while in the present study eggs were elongately oval and laid in clusters and hatching period was 4 days.

Rao *et al.* (1981) studied biology of exotic predator, *Chilocorus cacti* on sugar cane scale insect, *Melanaspis glomerata* Green. On an average the egg, grub and pupal periods were 5.94, 11.9 and 5.5 days respectively. The total life cycle from egg to adult was 22.3 days. In this species the eggs were brownish and cigar shaped and were laid singly on the scales infested canes. In all, 4 instars were reported, the 2nd and 3rd were voracious feeder while 4th instar was found feeding well in the early period and was slugish in the prepupal period. In the present study eggs were not cigar shaped and 2nd and 3rd instars were voracious feeders as like the above species. The egg, larval and pupal period averaged 4, 14 and 4 days respectively. Thus, life cycle was completed within 22 days.

Ghosh (1986) studied developmental periods in *Oenopia kriby* Mulsant, a predator of polyphagous sucking insect. He reported

12-14, 14-15 days and 6-7, 6-8 days as a periods of larval and pupal length on *Capitophorus formosartemisiae* and *Brevicoryne brassicae* respectively while, in the present study the larval and pupal periods were 14 and 4 days and the predator was reared on the aphid, *R. maidis*.

Diraviam and Virakatmath (1991) studied the biology of introduced lady bird beetle, *Curinus coeruleus* Mulsant under prevailing room temperature and relative humidity, wherein they reported 33.2, 40.3, 34.6 and 36 days for the development of a predator from egg to adult in May-June 1989, Jan-Feb. 1990, May-June 1990 and July-Aug. 1990 respectively. The body of *I. darbari* larva was arched on top and flat below, but in some tribes like coleopterini, however the dorsum is not uniformly arched and the head and the thorax are flatter than the abdomen.

The head of the larvae is frequently squarish with rounded corners in number of coccinellids, but in present form it was with strongly rounded lateral corners, but in the hyperaspini and platynaspini it is strongly transverse. However, in the tribe sernagiini the head is elongate (Kamiya, 1965). The sides of the head were usually rounded to greater extent in the tribe coccinellini or they are straight in platynaspini and chilocorini. In the present form the head was completely sclerotized as like the tribe coccinellini, chilocorini, etc. but the head is sclerotized only in some area in certain tribes like hyperaspini, stethorini and platynaspini, but in the scymnini the head is sclerotized only very slightly. The frontal suture was missing in the present form while it is well developed and "V" or "Y" shaped in certain tribes like epilachnini, chilocorini, etc. In the tribes scymnini, platynaspini and hyperaspini the frontal suture is absent sepecially in 3rd and 4th instar larvae. The antennae were 3 segmented in the present form, wherein first segment was short and wide and the 2nd segment usually rather longer, twice as long as first and then appears elongate. In most coccinellid larvae the 3rd antennal segment is small, slightly sclerotized and hardly noticeable while, in certain tribes it is dome like.

The mandibles were elongated, triangular in shape in the present species and showed apically bifid tooth. While in the tribe epilachnini the mandibles are equipped with 4 or 5 large teeth at the apex. In other cocoinellid larvae the mandibles have only one tooth (Hyperaspini, Scymnini, Brumus, etc.) or 2 teeth (Coccinellini, Chilocorini, Noviini, Coleopterini) at the apex. But in psylloborini and anistoticta, there is a row of smaller teeth and thick setae in tytthaspini below the apical tooth. In the tribe bulaeini the mandibles are very unusual, they have a 3rd small tooth in addition to 2 well developed apical teeth. This modification is due to the feeding behaviour because the larvae feed on pollen.

In the present form, the legs were elongate as like the tribe epilachnini, coccinellini, bulaeini and tytthaspini but the tribes like hyperaspini, stethorini and scymnini show short legs. The femur and tibiotarsus were cylindrical in the present species but in certain tribes, hyperaspini and coleopterini some times the femora are wide and almost quadrate.

The armature of both thorax and abdomen is rather variable and complex and hence has a considerable taxonomical significance. In general, the armature is in the form of setae, chalazae, tubercle, strumae, parascolus, sentus and scollus and scattered on thorax and abdomen. In the present form the dorsum of 8 abdominal segments showed a row of 4 strumae which form dorsolateral and dorsomedian groups. Usually a definite type of armature structure is a characteristic for each tribe thus the larvae of hyperspini and platynaspini are covered only with chalazae and setae, the scymnini and noviini have tubercles and tytthaspini and bulaeini have the strumae as like the psylloborini under which the present form comes. But the tribe coccinellini is exceptional in that, larvae of its different genera have variable structures.

There is consistency among coccinellid larvae in the morphological differences, between instars and changes occur in the colouration, porportions and armature of the body (Hodek, 1973). In the first and second instar larvae the strumae structures were weakly developed while in the 3rd and 4th instars they were well developed in the present form. However, change in body colouration was not observed in the instars. But, the proportion of the body was found changed with respect to the age.

Diraviam and Virakatmath (1991) made certain observations of biometry of immature stages of *C. coeruleus* wherein the length and width of eggs, 1st instars, 2nd instars, 3rd instars, 4th instars and pupae were ranged from 0.09 to 1 mm and 0.3 to 0.4 mm, 1.4 to 21. mm and 0.5 to 0.7 mm, 2.7 to 3.2 mm and 0.7 to 0.8 mm, 4.4 to 4.7 mm and 1.5 to 1.7 mm, 5.3 to 6.7 mm and 1.8 to 2.2. mm and 4.2 to 4.5 mm and 3.1 to 3.4 mm respectively while, in the present study the length and width ranged from 1 to 1.43 mm and 0.42 to 0.60 mm, 2.98 to 3.85 mm and 0.6 to 0.8 mm, 4.92 to 5.20 mm and 0.98 to 1.28 mm, 5.1 to 5.72 mm and 1. to 1.29 mm, 6.98 to 8.30 mm and 2.11 to 3.29 mm and 4.2 to 4.90 mm and 3.1 to 3.40 mm respectively. In *Scymnus* sp. the eggs averaged 0.45 mm and 0.2 mm, 1st instars 0.5 mm and 0.2 mm, 2nd instars 1.43 mm and 0.49 mm, 3rd instars 2.24 mm and 0.55 mm, 4th instars 3.2 mm an 0.98 mm and pupae 2.22 mm long and 1.58 mm broad (Patro and Behera, 1992). In *C. bijuqus* length and width of eggs, 1st instars, 2nd instars, 3rd instars, 4th instars and pupae averaged 0.93 ± 0.01 mm and 0.46 ± 0.01 mm, 1.15 ± 0.04 mm and 0.51 ± 0.02 mm, 2.34 ± 0.19 mm and 0.92 ± 0.90, 4.10 ± 0.17 mm and 2.16 ± 0.16 mm, 6.44 ± 0.20 mm and 2.94 ± 0.17 mm and 5.75 ± 0.16 mm and 3.70 ± 0.12 mm respectively (Rawat *et al.* 1992).

The length and width of larval body are frequently chosen as a characters which may help in determining the number of larval instars. The larval body length and breadth were increased as the age increases. The most sclerotized parts like head capsule,

Table 8 : Biometrical measurements* of the larval instars of *I. darbari*

Larval body structure	*First*	*Second*	*Third*	*Fourth*
		Larval instars		
I *Larval body*				
Length	3.50	5.06	5.69	8.00
Width	0.78	1.26	1.26	3.27
II *Head*				
Length	0.31	0.42	0.73	1.05
Width	0.42	0.54	0.84	1.09
III *Mandible*				
Length	0.12	0.14	0.16	0.18
Width	0.08	0.12	0.16	0.18
IV *Spiracle*				
Diameter	0.04	0.06	0.08	0.10
V *Vesicle*				
Length	0.42	0.25	0.52	0.94
Width	0.31	0.52	0.84	1.26

* Average measurement in millimeter of 20 individuals.

mandibles and spiracles were measued which does not showed increase in length and width as per the age but showed instar related growth. The results were tested by regression analysis and clearly indicates that there exists correlation between the age of larval instars and size. Statistical results are tabulated in tables 8 to 14. The present results are in agreement with the general principle of moulting.

Table 9 : Statistical linear regression and correlation of larval body length and age of larva in *I. darbarii.*

x = larval age,
y = larval body length.

u = log (x+1)	*v = log (1+y)*	u^2	v^2	*uv*
0.7781	0.6532	0.6055	0.4266	0.5782
1.0000	0.7041	1.0000	0.4958	0.7041
1.1139	0.7551	1.2408	0.5701	0.8411
1.1760	0.9030	1.3831	0.8155	1.0619
$\Sigma u = 4.068$	$\Sigma v = 3.0154$	$\Sigma u^2 = 4.2294$	$\Sigma v^2 = 2.308$	$\Sigma uv = 3.1853$

$\bar{u} = 1.017$, $\bar{v} = 0.7538$

$u - \bar{u} = buv\ (v - \bar{v})$

$\text{Log}\ (x+1) = -1.5467 + 3.4011 \log (1+y)$

$r = 2.09$

$r > 1$ ∴ correlation is perfect positive.

Table 10 : Statistical linear regression and correlation of larval body width and larval age in *I. darbarii.*

x = larval age,
y = larval body length.

u = log (x+1)	*v = log (1+y)*	u^2	v^2	*uv*
0.7781	0.2504	0.6055	0.0627	0.1948
1.0000	0.3541	1.0000	0.1253	0.3541
1.1139	0.3541	1.2408	0.1253	0.3944
1 1760	0.6304	1.3831	0.3974	1.7413
$\Sigma u = 4.068$	$\Sigma v = 1.589$	$\Sigma u^2 = 4.2294$	$\Sigma v^2 = 0.7107$	$\Sigma uv = 1.6846$

$\bar{u} = 1.017$, $\bar{v} = 0.3972$

$u - \bar{u} = buv\ (v - \bar{v})$

$\text{Log}\ (x+1) = 0.6743 + 0.8628 \log (1 + y)$

$r = 0.08014$

$r > 0$ ∴ correlation between larval age and body width is positive.

Table 11 : Statistical linear regression and correlation of head length and number of instar in *I. darbarii*

x = number of instar,
y = head length.

u = log (x+1)	*v = log (1+y)*	*u^2*	*v^2*	*uv*
0.3010	0.1172	0.0906	0.0137	0.0352
0.4771	0.1522	0.2276	0.0231	0.0726
0.6020	0.2380	0.3624	0.0566	0.1432
0.6989	0.3117	0.4885	0.0971	0.2178
$\Sigma u = 2.079$	$\Sigma v = 0.8191$	$\Sigma u^2 = 1.1691$	$\Sigma v^2 = 0.1905$	$\Sigma uv = 0.4688$

$\bar{u} = 0.5197$, $\bar{v} = 0.2047$

$u - \bar{u} = buv\,(v - \bar{v})$

$\text{Log}\,(x+1) = 0.1328 + 1.8903 \log\,(1 + y)$

$r = 0.9577$

$r > 0$ $\therefore$ correlation is positive.

Table 12 : Statistical linear regression and correlation of head width and number of instar in *I. darbarii*

x = number of instar,
y = head width

u = log (x+1)	*v = log (1+y)*	*u^2*	*v^2*	*uv*
0.3010	0.1522	0.0906	0.0231	0.0458
0.4771	0.1875	0.2276	0.0351	0.0894
0.6020	0.2648	0.3624	0.0701	0.1594
0.6989	0.3201	0.4885	0.1024	0.2237
$\Sigma u = 2.079$	$\Sigma v = 0.9246$	$\Sigma u^2 = 1.1691$	$\Sigma v^2 = 0.2307$	$\Sigma uv = 0.5183$

$\bar{u} = 0.5197$, $\bar{v} = 0.2311$

$u - \bar{u} = buv\,(v - \bar{v})$

$\text{Log}\,(x+1) = 0.0059 + 2.2235 \log\,(1 + y)$

$r = 0.9742$

$r > 0$ $\therefore$ correlation is positive.

Table 13 : Statistical linear regression and correlation of mandible length and number of instar in *I. darbarii*

x = number of instar,
y = mandible length

u = log (x+1)	*v = log (1+y)*	u^2	v^2	*uv*
0.3010	0.0492	0.0906	0.0024	0.0148
0.4771	0.0569	0.2276	0.0032	0.0271
0.6020	0.0644	0.3624	0.0041	0.0387
0.6989	0.0718	0.4885	0.0051	0.0501
$\Sigma u = 2.079$	$\Sigma v = 0.2423$	$\Sigma u^2 = 1.1691$	$\Sigma v^2 = 0.0148$	$\Sigma uv = 0.1307$

$\bar{u} = 0.5197$, $\bar{v} = 0.0605$

$u - \bar{u} = buv (v - \bar{v})$

$\text{Log} (x+1) = -0.9323 + 24 \log (1 + y)$

$r = 1.142$

$r > 1 \therefore$ correlation is perfect positive.

Table 14 : Statistical linear regression and correlation of mandible width and number of instar in *I. darbarii*

x = number of instar,
y = mandible length

u = log (x+1)	*v = log (1+y)*	u^2	v^2	*uv*
0.3010	0.0170	0.0906	0.0002	0.0051
0.4771	0.0492	0.2276	0.0024	0.0234
0.6020	0.0644	0.3624	0.0041	0.0387
0.6989	0.0718	0.4885	0.0051	0.0501
$\Sigma u = 2.079$	$\Sigma v = 0.2024$	$\Sigma u^2 = 1.1691$	$\Sigma v^2 = 0.0118$	$\Sigma uv = 0.1173$

$\bar{u} = 0.5197$, $\bar{v} = 0.0506$

$u - \bar{u} = buv (v - \bar{v})$

$\text{Log} (x+1) = 0.1339 + 7.625 \log (1 + y)$

$r = 1.025$

$r > 1 \therefore$ correlation is perfect positive.

PLATE - 25

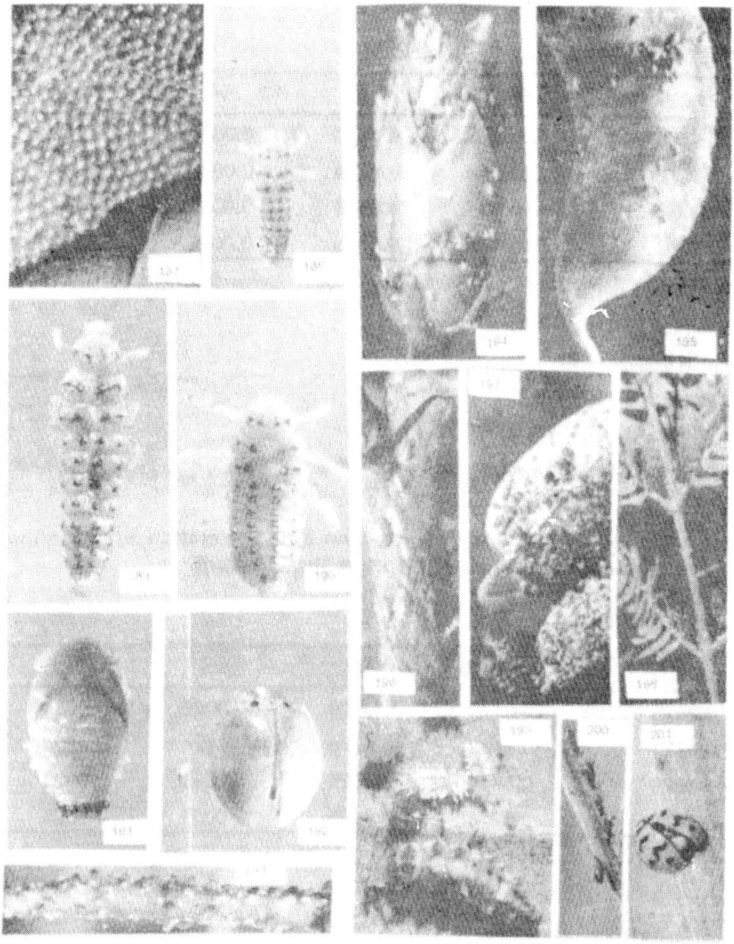

Biology of *I. darbarii* sp. nov.

Fig. 187 : Egg

Fig. 188 : 1st instar

Fig. 189 : 2nd instar

Fig. 190 : 4th instar

Fig. 191 : Pupa

Fig. 192 : Adult (female)

Fig. 193-195 : Aphids

Fig. 196 : Coccids

Fig. 197 : Aphids

Fig. 198 : Psyllids

Fig. 199 : *Menochilus* larva & aphids

Fig. 200 : Aphids

Fig. 201 : *Menochilus* adult

5

Predator Prey Relationships

Predators in general are always larger than the preys but exceptionally may be smaller than the prey. They required more than one preys for completion of their life cycle. They always kill their prey by hunting upon them. The life cycle stages like larva and adult, are free living in predators.

Texonomical and biological aspects of predators, in general, are studied, by several workers like Balduf (1933, 1935), Sweetman (1936, 58), Clausen (1940), Swan (1964), Mulsant (1866), Miyatake (1961), Kapur (1955, 63, 72), Bosale (1994), etc.

The insect predators are recorded from almost every order of the insects. However, the orders like Coleoptera, Neuroptera, Hemiptera, Diptera, Hymenoptera, Thysanoptera, etc. have major share. From order coleoptera, coccinelids alone contain 4000 described species. The crabids or ground-beetles are also important predators from this order. These beetles are characterized by having broad elytra and black colouration or brillient metalic blue/green/purple colouration, narrow pronotum and long legs. The larvae shows two spike like processes at the posterior and they are flattened and slender. The tiger beetle, *Pheropsophus* sp. feed on rhinocerous grubs, *Oryctes rhinocerous*. Some rove beetles are also predaceous. They feed on eggs of certain Lepidopterous pests and Dipterous larvae.

Neuropteran chrysopids are greenish and slender bodied, with golden eyes and lace like wings. *Chrysopa carnea* feed on the eggs of cotton bollworms. While, Hemerobids feed on several sucking type of insects like aphids, mealy bugs, adelgids, white flies etc.

From hemiptera several species are reported as predatory bugs, eg. Pentatomids, Mirids, Ruduviids, Corids etc. Mirids are velvety in appearance and measures about 2.5-6 mm in body length. They are characterized by having 4 segmented antenna and rostrum and no oceli. *Rhinocoris* sp. and *Coranus* sp. feed on tassar larvae. Many Ruduviids feed on pests like aphids, jassids, sawfly larvae and some lepidopteran larvae.

Syrphids from order Diptera are prominant predators which feed on the insects at their larval stage. The syrphids are 3-25 mm long, borad or slender, polished black or metalic coloured. The syrphid larvae are leg less and elongated with sucking mandibles.

The Noctuids like *Eublema* sp. from order lepidoptera, feed on mealy bugs and soft scales. Similarly, larvae of Lycaenids and Epipyropids of lepidoptera also feed on soft scales of hemiptera order.

The predators may have either chewing type of mouth parts or sucking type. They may be specific or monophagus as like *Rodolia cardinalis* which feed on *Icerya purchasi*. They may be oligophagus, in which their preys have the range within one family only, ex : syrphids on Aphididae (Aphids). The polyphagus predators have prey range with a wide variety of families. The examples of polyphagus predators include, praying mantids, dragon flies, ground beetles, ants, etc. The predators may have appropriate physiological or morphological adaptations.

The predators can be sampled by several methods. Some important methods are given below : simple sunction trap, air layer, jarring or plant checking, malaise traps, mite brushing machine, sucking apparatus, drop trap, sweep nets, cylinder or chamber, D-Vac R, digging squar meter or squar foot, pitfall traping etc.

Predator Population Influencing Factors

The population of predators can be influenced by the basic needs of an animal i.e.

1. Food
2. Shelter and
3. Mate

The predator population can be influenced by :

1. *Biotic factors* : Such as natural enemies and diseases etc. Natural enemies are from parasites/parasitoids/ pathogens etc. The coccinilids and *Chysopa* sp. are parasitized by hymenopterous wasps.
2. *Abiotic factors* : Such as wheather (temp., humidity, rainfall, light etc.) topographic factors, habitat, pesticides etc. are directly or indirectly related to the survival of predators.

   The efficiency of predator species is dependent on prey density, predator density, characteristics of the prey and predator and environment. Likely, the functional responses of predators are dependant on quality of food, imigration, emigration, reproductive capacity, mortality, searching, discovery, handling time, digestion, hunger, prey defences and competition etc.

Sources of Predators

The predators in general are more abundant where there is good source for their preys. On an annual crops, orchards, in persistance agroecosystem, etc. the predators would be more abundant.

Legume crops are important source for many predators, wheat, jowar, barley, etc. are good sources for predators like *Chrysopa* sp., *Geocoris* sp., Lady bird beetles, syrphids etc. Mulberry is good source for predators like preying mantids and coccinelids etc. While, corn is good source for Nabids which feed on aphids. The aphids are associated with more than 18 families of plant kingdom. However, they are more abundant on many plants of the family polygoneaceae hence, these plants are good source for coccinelids and other predators. Similarly, the family composite is also good source for many predators. Vegetable crops, groundnut, safflower, many pulse crops get attacked by aphids and other homopterous insects hence these crops are also good source of predators like *Crysopa*, preying mantids, many coccinelids and plant bugs, etc. The cotton is very good source of predators like *Menochilus sexmaculatus*, *Coccinella* sp., *Scymnus* sp., *Chrysopa* sp., and several species of spider. The availability of predator population is dependant on the abundance of prey, abundance of prey plants, season, climate and ethology of the predators & preys. Many lady bird beetles gather on the plants of hilly region in summer for mating and for their food.

In India, there is need to have a detailed survey of predators of pests and the sources of the predators. There is endless scope to do research on the various aspects of pest predators in India.

6

Predator Prey Index and Biocontrol Programmes

Very large number of predators are reported on insect pests from different parts of the world. The list of important predaceous species and the pests they attack are listed in the following table. The pest control programmes adopted by using the predaceous species are also listed in the Table a and b.

Insect pest predators found in India

Sr. No.	*Pest species*	*Predator species*	*Pest stage attacked*
I : *Hemipteous pests :*			
1.	The sorghum shoot bug	Ladybird beetles	L
	Peregrinus maidis Ashmead	*Coccinella septumpunctata* L.	L
2.	The sorghum aphid,	Coccinelid beetle,	L
	Rhopalosiphum maidis	*Menochilus sexmaculatus* Fab.	
	(Fitch)	*M. shivajensis*	
		M. kharipi	
3.	Aphid - *Dactynotus*	*Brumoides kolhapurensis*	L+A
	compsitae Theobald	*Chilochorus suryaphuli*	L+A
		Coelophora orientalis	L+A
		Menochilus tuljapurensis	L+A
4.	*R. maidis*	*Illeis darbari*	L+A
5.	*D. compositae*	*Scymnus indica*	L+A
6.	*Pseudoletia separata*	*S. indica*	E+L
7.	*Aphis gossypi* (Glov.)	*Deraecoris indicus*	N+A
		Coccinella yamini	N+A
		Hoverfly larvae	N
		lacewing larvae	N+A
8.	The mustard aphid,	Coccinelods -	
	Lipaphis erysimi Kalt.	*Coccinella septumpunctata* L.	N+A
		Menochiles sexmaculata (Fab.)	N+A

Contd...

Sr. No.	*Pest species*	*Predator species*	*Pest stage attacked*
9.	*Myzus persicae* (Sulzer)	*Deraecoris indicus*	N+A
		Illeis darbari	N
10.	The groundnut aphid,		
	Aphis carccivora Koch.	*Brumoides* sp.	N+A
		Coccinella transversalis	N
		C. bhuimungi	N+A
		Harmonia soyabini	N+A
		Menochiles sexmaculatus Fab.	N
	Syrphid,	*I shiodan javana* (Widdmann)	
11.	Aphid	*Orius* sp.	N+A
12.	The citrus psylla,	Coccinelids,	
	Diaphorina citri K.	*Coccinella septumpunctata* L.	
		C. rependa	
		Chilomenes sexmaculata	N+A
		Chilocorus nigrita	
		Brumus suturalis	
		Chrysoperala carnea (N.)	
13.	The citrus white fly,	Ladybird beetles	
	Dialeurodes citri Asmead	*Brumus suturalis* Fab.	N
		Veronia cardoni Weire	
		Cryptognatha flavescena Motsch.	N
		Thirps - *Aleurothrips fasciatipennis*	
		Lacewing - *Chrysopa* sp.	
14.	The citrus red scale,	Coccinelid -	
	Aonidiella auranti	*Chilocorus nigritus*	N
	(Maskell)	*Scymnus quadrillum*	N+A
15.	The coconut scale,	Coccinelids -	
	Aspidoitus destructor Sign.	*Azay trinitalis* Mshl.	
		Chilocorus sp.	N+A
		Pentilia castanea Muls.	
		Pharoscymnus grimeti Muls.	
16.	The cottony cushion scale	*Rodolia cardinalis*	N+A
	Icerya purchasi (Maskell)		
17.	The softscale,	The ants	
	Coccus viridis (Green)	*Oecophylla smaragdina* Fab.	
		Camponotus sp.	N
		Cremastogaster sp.	
18.	The soft scale,	*Spalgis epius* Westwood	N
	Pulvinaria sp.	*Spalgis* sp.	
19.	The psyllid,	Coccinelid -	
	Heteropsylla	*Coccinella bhuimungi*	N+A
	cubana Crowford 7		
20.	The cotton white fly	Coccinelid - *Brums* sp.	N
	Bemisia tabaci Genn.	Lacewing - *Chrysoperla* sp.	N+A

Contd...

Sr. No.	Pest species	Predator species	Pest stage attacked
21.	The mulberry whitefly *Aleurodicus dispursus* Russel	Ladybird beetles *Menochillus sexmaculatus* *Coccinella septumpunctata* *Cryptolaemus montrouzieri* *Scymnus* sp.	N
22.	The mango mealy bug *Drosicha mangiferae* Green	Coccinelid *Rodolia cardinalis* Mulsant Lace wing, *Chrysopa sclestes* Banks	N
23.	The jute mealy bug *Phenacoccus hirsutus* Cr.	Coccinelid, *Scymnus pallidicollis*	E,N+A
24.	The sugarcane mealy bug *Saccharicoccus sacchari* (Cock.)	Coccinelid, *Scymnus coccivora* R. *S. andrewsi* Sie. *S. nubilis* Muls. *Hyperespis trilinineala* Muls. *Pharoscymnus grimeti* Muls. Ladybird beetles -	N+A N
25.	The Paddy mealy bug *Ripersia oryzae*	Lady bird beetle	N
26.	*The lac insect, *Karria lacca* (Ker.)	Noctuid larva, *Eublemma* sp.	N
27.	The chermids	*Hemerobius adelgivorus* Kim. *Sympherobius* sp.	N
28.	The jassid, *Ampoasca devastans*	*Illeis sathei*	N
29.	The cotton jassid *Amrasca devastans* (Distant)	Lace wing, *Chrysopa* sp, Spider, *Distina albida* L.	N+A
30.	The pyrilla, *Pryilla purpusilla* Walker	Coccinelid, *Brumus suturalis* Fab. *Chilomenes sexmaculatus* Fab. *Coccinella septumpunctata* Linn. Crysopods : some species	E+N
31.	Red cotton bug, *Dysdercus cingulatus* Fab.	Pyrrhocoridbug, *Atilochus cocaueberti* F. Reduviid bug, *Harpactor costalis* Rev.	N
32.	The dusky cotton bug, *Oxycarenus laetus* kirby	Anthocorid bug, *Triphleps tantilus* Motsch.	E+N
33.	The brown leaf hopper, *Nilaparvata lugens* (Stal)	Mird bug, black ants *Camponotus* sp., Lady bird beetles.	N

Contd...

Sr. No.	Pest species	Predator species	Pest stage attacked
II : *Orthropteran pests :*			
34.	Deccan grasshopper - *Colemania spenarioides*	Blister beetle, *Zonabris* sp. Bombylid fly, *systaechus* sp.	E
III : *Lepidopteran pests :*			
35.	The gram pod borer *Helicoverpa armigera* (Hubn.)	Coccinelid, *Menochilus* sp. Lacewing, *Chrysopa scelestes* Blanks, *Chrysoperla carnea* (Steph.) *Rhynocoris luscipes* Fab., *R. marginatus* Fab. *Sycanus indagaster* Stal. *Coranus spiniscutis* Reut, *Geocoris* sp., *Nabis* sp,	E+N
	Geocoris sp. *Nabis* sp.	Reduviids, Spiders *Thomisus* sp. *Neosiona* sp. Crows - *Corvus* spp.	L
36.	The codling moth *Cydia pomonella* (L.)	Indian house sparrow, *Passer domesticus* indicus	
37.	The caster slug caterpillar, *Parasa lepida* C.	Pyralid caterpillar, *Phycita dentilinella* H.	
38.	The pink boll worm, *Pectinophora gossypiella* (Sauders)	Anthocorid bug, *Triphles tantilus* Mot.	E+L
39.	The indigo caterpillar, *Spodoptera exigua* (Hubn.)	Pselaphid beetle, *Euplecuts* sp.	E+L
40.	*Nephantis serinopa*	*Parena laticincta* Bates	
41.	*The mulberry silk worm, *Bombyx mori* L.	Desmestid beetles, *Dermestis cadverinus* Preying mantid- *Hierodula bipapilla* Reduviid- *Sycanus collaris* Stink bug - *Canthecona furcellata* Ant, *Oecophylla smaragdina*	L
42.	**Tassar silk worm, Antheraea mylitta***	*Preying mantid, Hierodula pipapilla* Reduviid, *Sycanus collaris* Stink bug, *Canthecona furcellata* Ant, *Oecophylla smaragdina*	L L L
IV : *Coleopteran pests :*			
43.	The rhinocerous beetle *Oryctes rhinocerous* Linn.	The carabid, *Pheropsophus hilaris* *Sobrinus* D. *Pheropsophus* sp. *Platymeris laevicollis* D.	L L L L

Contd...

Sr. No.	Pest species	Predator species	Pest stage attacked
		Others :	
		Frog, toads, birds, rats and quirrels	L
44.	The white arub, *Holotrichia consanguinea* (Blan.)	Indian toad, *Bufo melanostictus*	A
		The wall lizard, *Giecko gecko*	

*Useful insects
E : Egg
L : Larva
N : Nymph
A : Adult

Table a : Biological pest control programmes conducted by using predators

Sr. No.	Pest species	Predator species	Country
1.	The coconut scale, *Aspidoitus destructor* Sign.	West Indian Ladybird beetle, *Cryptognatha nodiceps* Marshall	Fiji
2.	The blue gum scale, *Eriococcus coriaceus* Mask.	Rhizobius ventralis (Eric.)	New Zealand
3.	The cottony cusion scale I *cerya purchasi* Eric.	Ladybird beetle, *Rodolia cardinalis* Mulsant	California India
4.	The mealy bugs	*Cryptolaemus montrouzieri* Mulsant	India and many other countries
5.	The Rhinocerous beetle *Oryctes rhinocerous*	*Platymeris laevicollis* Dist.	India
6.	The cotton boll worms	*Chrysopa carnea* Steph.	India, USA
7.	*Pseudococcus maritimum* (Eh.)	*Chrysopa carnea*	India
8.	*Heliothis* sp.	*Chysopa carnea*	USA

Table b : Biological pest control programme

Sr. No.	*Pest species*	*Predator species*	*Country*
9.	*Thanatus flavidus* (spider)	Bed bug *Cimex* sp.	Greek
10.	*Biscirus lapidarius* (Kramer)	*Smithurus viridis* (L.) (Collumollan)	Australia
11.	*Chlorohydra viridissima* (Coelentrate)	*Culex tarsalis* (Lud.)	USA
12.	*Dugesia dorotocephala* (Woodworth) (Playtyhelminthes)	Mosquitoes, midges	California,
13.	*Gambusia affinis* (B. & G.) (Fish.)	Mosquitoes, *Anopheles* sp.	Hawai and several countries
14.	*Bufo marinus* (L.) (Amphibian)	Molecriket, june beetle & cockroaches	Puerto Rico
15.	Indian mynah bird, *Acridotherus tristis* L.	Red locust, *Nomadacris septumfasciata* Ser.	Mauritius
16.	Masked shrew	Larch fly	Canada
17.	Moles & Shrews	Pine saw fly & larch sawfly	Canada

Summary

Insect pest control has become the most pressing problem in recent years. The pesticides solve the problem upto certain extent but, lead to many serious problems like pest resistance, secondary pest out break, pest resurgence, health hazards, air & water pollution, etc. Biological control is living weapon over chemical control. Pradators are widely used in biological pest cotnrol in the world as a biocontrol agents. The coccinelid beetels have tremendous potential for utilization in biological pest control. Any advance in knowledge of taxonomy, biology, ethology, predator prey index and utility of predators is therefore, of potential practical value in pest management.

Introduction of the text contain agricultural potential and achievements in the field, destructive insect pests and their control and importance and activities of the biological control of India. In second chapter, methods of collection, preservation and rearing of predators are described. Third chapter includes the taxonomical description of 19 species of coccinelids (Coleoptera : Coccinelidae) of 10 genera, *Brumoides* chapin, *Chilochorus* Leach, *Coccinella* Linneaus, *Coelophora* Mulsant, *Hormonia* Mulsant, *Menochilus* Timberlake, *Vernia* Mulsant, *Illeis* Mulsant, *Horniolus* Weise, *Scymnus* Kugelann belonging to 4 tribes *Chilocorini, Coccinellini, Psylloborini* and *Scymnini* of sub families chilochorinae, Coccinelinae and Scymninae. Life cycle and biometry of immature stages of *Illeis darbari* are important aspects of fourth chapter. In fifth chapter, predator pray relationships are described. Sixth chapter is devoted for insect pest predator index and the control programmes adopted using insect and noninsect predators in India and abroad. Seventh chapter refers to summary and eighth bibliography.

In India, no such type of monograph/book is published on Indian coccinelids, as it is very important group of insects utilized in biological pest control. This would be very significant contribution on predatory insects of pest species in India and will be helpful for designing biocontrol programmes of pests.

Bibliography

Note : W. L. indicates the number of the periodical in the "world list of scientific periodicals" 4th Edn. (1963-65) edited by P. Brown and C.B. Strawon. This number is given at the first citation of the periodical only.

Aiyar, T.V.R., 1924. An undescribed Coccinellid beetle of economic importance. *J. Bombay Nat. Hist. Soc.*, 30 : 491-483. (W. L. 25676).

Alexander, R.D., 1961. Aggressiveness, territoriality and sexual behaviour in field crickets (Orthoptera : Gryllidae). *Behaviour*, 17: 130-223.

Anderson, W.H., 1936. A comparative study of the labium of Coleopterous larvae. *Smithson. misc. Coll.*, 95 (13) : 1-29.

Anonymous, 1987. Districtwise General Statistical Information of Agriculture Department (1965-1987) 2 : 210-213.

Arnett, R.H., 1963. The beetles of the United States. PP. 1112. Washington.

Arrow, G.J., 1925. The Fauna of British India, including Ceylon and Burma Coleoptera-Clavicornia, Erotylidae, Languridae and Endomychidae, PP 416 I Col. P1., London.

Atkins, M.D., 1980. Introduction to insect behaviour, PP - 1-237. Macmilian Publishing Co., Inc. New York.

Balduf, W.V., 1935. The biononics of entomophagous Coleoptera (13. Coccinellidae - lady beetle). PP 220. John S. Swift Co. Inc., Chcago, New York.

Banks, C.J., 1955. An ecological study of Coccinellidae associated with *Aphis fabae* Scop. on *Vicla faba*. *Bull. ent. Res.*, 46, 561-587. (W.L., 10184).

Bennam, B.R., Mugleton, J., 1970. Studies on the ecology of *Coccinella undecimpunctata* Linn. (Coleoptera : Cocclinellidae). *Entomologist*, 153-170.

Benkevich, V.I., 1958. Biology of *Coccinella septempunctata* Uchem. Zap. Orekh. *Zuey Pedag. Inst.*, 11 : 127-133.

Binaghi, G., 1941. Larve pupe di Chilocor ini. *Memorie Soc. ent. Ital.*, 20 : 10-36.

Blackman, R.L., 1965. Studies on specificity in Coccinellidae. *Ann. appl. Biol.*, 56 : 336-338.

Blackman R.L., 1967. Selection of aphid prey *Adalia bipunctata* L. and *Coccinella 7-punctata* L. *Ann. appl. Biol.*, 59 : 331-338.

Bodenheimer, F.S., 1943. Studies on the life history and ecology of Coccinellidae I. The life history of *Coccinella 7-punctata* L. in four different Zoogeographical regions. *Bull. Soc. Fauad. I. Ent.*, 27 : 1-28.

Booth, R.G., 1991. A new genus and species of Coccinellid (Coleoptera : Coccinellidae) predator on *Rostrococcus* spp. *Bull. ent. Res.*, 81 : 147-150.

Boving, A.G., 1917. A generic synopsis of the Coccinellid larvae in the United National Museum with a description of the larva of *Hyperapis binotata* Say. *Proc. U.S. natn. Mus.*, 51 : 621-650.

Boving, A.G., 1933. Description of the larva of *Decadiomus pactus* Chapin (Scymnini : Coccinellidae). *Proc. Biol. Soc. Washington.* 46(1) : 101-104.

Brewbaker, James L., 1987. Species in the genus *Leucaena*, Leucaena Research Reports. *Proc. work. Bio. Genetic control strat. leucaena psyllid.* 7(2), : 6-20.

Brown, W.J., 1962. A revision of the forms of *Coccinella* L. Occurring in America north of Mexico (Coleoptera : Coccinellidae). *Can.Ent.*, 94 : 785-808.

Brown. W.J. and R.de Ruette., 1962. An annotated list of the Hippodamiini of Northern America with a key to genera (Coleoptera : Coccinellidae). *Can. Ent.*, 94 : 643-652 (W.L. 13141).

Buck, J.B., 1948. The anatomy and physiology of the light organ in fireflies. In Bioluminescence. *Ann. N. Y. Acad. Sci.*, 49 : 397-482.

Butt, F.H., 1951. Feeding habits and mechanism of the Mexican Bean beetle. *Mem. Cornell Univ. Agri. Exp. Sta.* 306 (1) : 1-31.

Canepari, C., 1986. On some coccinellids of Northern India and Nepal in the Geneva Museum of Natural History (Coleoptera, Coccinellidae). *Reveu Suisse Zool.*, 93 : 21-36.

Cardona, C. and Oatman, E.R., 1971. Biology of *Apanteles dignus* (Hymenoptera : Braconidae), a primary parasite of the tomato pinworm. *Ann. ent. Soc. Am.*, 5 : 996 -1007 (W.L. 3145).

Casey, T.L., 1899. A revision of the American Coccinellidae. *J.N.Y. Ent. Soc.*, 7 : 71-169.

Casey, T.L., 1908. Notes on the Coccinellidae. *Can. Ent.*, 40 : 393-421.

Chapin, E.A., 1962 *Pseudoscymnus*, a new genus of Asiatic Scymnini (Coleoptera : Coccinellidae). *Psyche*, 69 : 50-51.

Chapin, E.A., 1965. Coleoptera : Coccinellidae. *Ins. Micronesia.*, 16 : 189-254.

Chapin, E.A., 1965. The genera of the Chilocorini (Coleoptera : Coccinellidae). *Bull. Mus. Comp. Zool.*, 133 : 227-271.

Chapin, J.B., 1971. The Coccinellidae of Louisiana (Insecta : Coleoptera) Doctrola dissertation. Luisiana State University.

Chapin, J.B., 1973. New species of lady beetles in the genera *Scymnus* and *Mulsantina* (Coleoptera : Coccinellidae). *Ann. ent. Soc. Am.* 66 : 1071-1073.

Chapin, J.B., 1974. The Coccinellidae of Louisiana (Insecta : Coleoptera) *Agri.Exper. Sta.* Doyle Chambers, *Bull.* 682, pp. 1-87.

Chapius, M.F., 1876. In Lacordaire, Natural des Insects et Genera des Coleopteres, Vol. 12, Famillies des Erotyliens, des Endominchides, et des Coccinellides Paris. (In French).

Chapman, R.F., 1969. The insects : structure and Function. American Elsevier Publishing Co. Inc., New York.

Chumakova, B.M., 1962. Experiments in rearing of predatory beetle *Cryptolaemus montrouzieri* Muls. on an artificial diet. *Zesz. Probl. Postep Nank roln.*, 35 : 195-200

Clausen, C.P., 1940. Entomophagus insects. PP 688. McGraw-Hill, New York-London.

Clements, F.E., Shelford, U.E., 1939. Bio-Ecology pp 425. Wiley Inc. New York.

Colburn, R., Asquith, D., 1970. A cage used to study the finding of a host by the lady bird beetle *Stethorus punctum. J. econ. Ent.*, 63 : 1376- 1377 (W.L. 22997).

Coppel, H.C. And Mertins, J.W., 1977. Biological Insect Pest Suppression. *Adv. Series Agri. Sci.*, 4, 1-301.

Corbet, P.S., Longfield, C. and Moore, N.W., 1960. Dragonfiles. Collins Clear-Type Press, London.

Corbet, S.A., 1971. Mandibular gland secretion of larvae of the flour moth, *Anagasta knehniella* contains an epidrietic pheromone and elicits oviposition movements in a hymenopteran parasite. *Nature*, 232:481-84. (W.L. 34258).

Costa, A., 1849. Fauna del Regno di Napoli. PP. 112. *Coleoterri, Napoli, Sautto.*

Crotch, G.R., 1873. Revision of the Coccinellidae of the United States. *Trans. Amer. entomol Soc.*, 4 : 363-382.

Crotch, G.R., 1874. A Revision of the Coleopterous family Coccineillidae. PP 311., London (I, II)

Crowson, R.A., 1960. The Phylogeny of Coleoptera. *A.Rev. Ent.*, 5(1), 111-134 (W.L. 3436)

Debach, Paul, 1964. Biological Control of insect pests and weeds. PP:1-843. Champan and Hall Ltd., London.

Diraviam, J. and Virakatmath, C.A., 1991. Population Dynamics of the introduced lady bird beetle, *Curinus coeruleus* Mulsant in relation to its psyllid prey *Heteropsylla cubana* Crawford in Bangalore. *J. Biol. control*, 4(2) : 99-104.

Diraviam, J. and Viraktamath, C.A., 1991. Biology of the introduced lady bird beetle, *Curinus coeruleus* Mulsant (Coleoptera:Coccinellidae). *J. Biol. Control*, 5(1), 14-17.

Diraviam, J. and Viraktamath, C.A., 1992. Laboratory studies on the host range of *Curinus coeruleus* Mulsant an Exotic predator of the psyllid. *J. Biol. Control*, 6(1), 42-43.

Dixon, A.F.G., 1959. An Experimental study of the searching behaviour of the predatory Coccinellid beetle *Adalia decimpunctata* (L.) *J. Anim. Ecol.*, 28 : 259-281 (W.L. 2559).

Dobzhansky, T., 1924. Dir Weiblichen Gener at jonsorgane der Coccitnelliden als Artmerkmal hetrachtet (Col). *Ent. Mitt.*, 13 : 18-27 (In German).

Dobzhanskhy, T., 1924. Die palaarktischen Arten der Gattung *Coccinellia* L. *Rev. Russ. Ent.*, 20 : 16-32.

Dowell, R.V. and Horn, D.J., 1975. Mating behaviour of *Bathyplectes cruculionis* (Hym. : Ichneumonidae), a parasitoid of the Alfalfa weevil, *Hypera postica* (Col. : Curculionidae) *Entomophaga*, 20:271-273 (W.L. 18271).

Downes, J.A., 1958. Assembly and Mating in the biting Nematocera. *Proc. 10th Int. Congr. Entomol., Montreal* (1956) 2:425-434.

Dunn, J.A., 1949. The parasites and predators of potato aphids. *Bull. ent. Res.*, 40 : 97-122.

Dutt, G.R., 1927. Aphids and lady bird beetles. *Agric. India* 22(4) 291-292.

Dyadechka, N.P., 1954. Coccinellids of the Ukrainian SSR. pp 156. Keiv (in Russia).

Eastip, V.F., Pope, R.D., 1969. Notes on the biology of some British Coccinellidae, *Entomologist*, 102 : 162-164.

Economic Survey of Maharashtra, 1991-92. pp 13-20. Directorate Statistical Department Maharashtra State.

Edwards, J.G., 1957. Entomology above timberline. II. The attraction lady bird beetles to mountain tops. *Coleopterists Bull.* 2, 41-46.

Ellingsen, I.J., 1969. Fecundity, aphid consumption and survival of the aphid predator *Adalia bipunctata* L. (Col. : Coccinellidae). *Norsk ent. Tidssk.*, 16, 91-95.

Fabricius. J.C., 1775. Systems Entomologiae pp 832. Lipsiae.

Fink, D.E., 1926. The biology of *Macrocentrus imvora* Rohwer, an important parasite of the strawberry leafroller (*Anylis comptana Forehl.*) *J. Agr. Res.*, 32:1121-1134.

Gage, J.H., 1920. Larvae of Coccinellidae. *Illinois. Biol. Mon.* 6(4) : 232-294.

Ganglbaver, L., 1899. Die Kafer Von Mitteleuropa Band 3, PP 1044. Wein. (In German).

Geyer, J.W.C., 1947. A study of the biology and ecology of *Exochmus flavipes* Thunb. (Coccinellidae : Coleoptera). *J. ent. Soc. South Africa* 9 : 219-234.

Ghani, M.A., 1962. A note on the identity of some species of genus *Ballia* (Coleoptera : Coccinellidae). *Proc. R. Ent. Soc. London (B).* 31 : 7-8.

Ghorpade, K.D., 1974. Description a new *Cryptoqonus* Mulsant from Bangalore Southern Indian. (Coleoptera : Coccinellidae). *Oriental Ins.*, 8 : 55-60.

Ghorpade, K.D., 1977. A new species of *Pseudoscymnus* (Coleoptera : Coccinellidae) predaceous on Coconut scale, in Peninsular India. *J. Bombay Nat. Hist.*, 11 : 465-469.

Ghosh, D., Debnath, N. and Chakarbarti, S., 1986. Development of three species of *Oenopia* (Coleoptera : Coccinellidae) predaceous on aphids of Garhwal Himalaya. III Oriental Entomolgy Symposium, Feb. 1984. Trivandrum proceedings, March 1986 pp. 173-176.

Gordon, R.D., 1978. West Indian Coccinellidae II (Coleoptera) : some scale predators with keys to genera and species. *Coleopterists Bull.*, 32 : 205-218.

Gordon, R.D. and Chapin, E.A., 1983. A revision of the New World species of *Stethorus* Weise (Coleoptera : Coccinellidae) *Trans. Am. ent. Soc.*, 109 : 229-276.

Gordon, R.D., Pakluk, J., Slipinksi, S.A., 1989 Carinodulini, a new tribe of Sticholotidinae based upon a new genus and species from Mexico (Coleoptera, Coccinellidae). *Coleopterists Bull.*, 43 : 359-364.

Hagen, K.S., 1962. Biology and ecology of predaceous Coccinellidae. *An. Rev. Ent.*, 7 : 289-236 (I).

Hariri, G.A., 1966. Laboratory studies on the reproduction of *Adalia bipunctata* (Coleoptera: Coccinellidae). *Entomologia Exp. Appl.*, 9 : 200-204.

Haseena Bhaskar, 1992. Systematic and Ecological studies on predaceous Coccinellid beetles (Coleoptera : Coccinellidae). M.Sc. (thesis) Dept. of Agricultural Entomology U.A.S. Bangalore pp. : 1-160.

Hiroyiki Kamiya, 1966. On the Coccinellidae attacking the scale insects and mites in Japan and Ryukyus. *Mushi.* 39(7) : 65-93. (W.L. 33872).

Hodek, I., 1966. Ecology of aphidophagous insects. *Proc. Symp. Liblice near Praque*, 1965, PP 360. Academia, Praha. Dr. W. Junk. Haag.

Hodek, I., 1967. Bionomics and ecology of predaceous Coccinellidae. *An. Rev. Ent.*, *12* : 79-104.

Hodek, I., 1973. Biology of Coccinellidae. PP 260. Academia, Prague.

Horn, G.H., 1895. Studies in Coccinellidae. *Trans. Amer. Entomol. Soc.*, 22 : 81-114.

Iwata, K., 1932. On the biology of two large lady birds in Japan. *Trans. Kanasi Entomol. Soc. Osaka*, 3 : 13-26.

Johnson. R.H., 1907. Economic notes on aphids and Coccinellids. *Ent. Nes. Philadelphia*, 18 : 171-174.

Kaddon, I.K., 1960. The feeding behaviour of *Hippodamia 5 signata* (Kriby) larvae. *Unvi. Calif. Publs. Ent.*, 16 : 181-230.

Kamiya, H., 1961. A revision of the tribe Scymnini from Japan and Loochoos (Coleoptera : Coccinellidae). *J. Fac. Agr. Kyushu Univ.*, 11 : 275-302.

Kamiya, H., 1965. Comparative morphology of larvae of the Japanese Coccinellidae, with special reference to the tribal phylogeny of the family (Coleoptera). *Mem. Fac. lib. Fukui Univ.*, (2), 14 : 83-100.

Kapur, A.P., 1947. A revision of the tribe Aspidimerini Weise (Coleoptera : Coccinellidae).

Kapur, A.P., 1948a, On the old world species of the genus *Stethorus* Weise (Coleoptera : Coccinellidae). *Bull. ent. Res.*, 39: 297-320.

Kapur, A.P., 1948b. On the Indian Species of *Rodolia* Mulsant (Coleoptera : Coccinellidae). *Bull. ent. Res.*, 38 : 531-538.

Kap·ır, A.P., 1950. The biology and external morphology of the larvae of Epilacninae. *Bull. ent. Res.*, 41 : 161-208.

Kapur, A.P., 1954. Systematic and biological note on the San Jose scale in Kashmir with description of a new species (Coleoptera : Coccinellidae). *Rec. Indian Mus.*, 52 : 257-274.

Kapur, A.P., 1955. Coccinellidae of Nepal. *Rec. Indian Mus.*, 53 : 309-338.

Kapur, A.P., and Rao, S.V., 1962. A new species of ladybird beetle [*Adalia simmondsi* sp. n.] (Col; Coccinellidae) feeding on *Adelges* [*Chermes*] spp. (Hem.; Adelgedae) on silver fir [*Abies* sp.] in North-Western India. *Entomophaga*, 7 : 47-51.

Kapur, A.P., 1963. The Coccinellidae of the third Mount Everest Expedition, 1924 (Coleoptera). *Bull. Brit. Mus. (nat. Hist.), Ent.* 14 : 1-48.

Kapur, A.P., 1967. The Coccinellidae (Coleoptera) of the Andamans. *Proc. Nat. Inst. Sci. India* 32(B); 148-189.

Kapur, A.P., 1969. On some coccinellidae of the tribe Telsimiini with description of new species from India. *Bull. Syst. Zool., Calcutta*, 1(2) : 45-56.

Kapur, A.P., 1972. The Coccinellide (Coleoptera) of Goa. *Rec. Zool Survey India*, 66 : 309-320.

Kapur, A.P. and Munshi, S., 1970. A new species of *Scymnus (Pullus)* Mulsant *S. (P) nymphaeus* sp. nov. (Coleoptera : Coccinellidae) Predaceous on the aphid *Rhopalosiphum nymphaeae* (Linn.) in Calcutta, India. *Rec. Zool. Survey India*, 62 : 1-4.

Katakura, H., 1985. Sperm transfer in the Potato lady bird *Hensoepilachna vigintioctomaculata* (Coleoptera : Coccinelidae : Epilachninae). *Kontyu*, 53 : 652-657.

Katyal and Chadha 1987. Vegetable growing in India. pp. : 1-149.

Kehat, M., 1968. The feeding behaviour of *Pharoscymnus numidieus* (Coccinellidae), predator of the date palm scale, *Parlatria blenchardi. Entomologia exp. Appl.*, 11 : 30-42.

Klausnitzer, B., 1969. Zur Kenntnis derlarve Von *Lithophilus conatus* (Panzer) (Col., Cocecinellidae). *Ent. Nachr.*, 13 : 33-38.

Koide, T., 1962. Observations on the feeding habit of the larva of *Coccinella septumpunctata bruckii* Mulsant. The feeding behaviour and number of prey fed under different temperatures. *Kontyu*, 30 : 236-241.

Korschefsky, R., 1931. Coccinellidae I. pp 224. Coleopterorum Catalogues, Pars 118, Berlin.

Korschefsky, R., 1932. Coleopterorum Catalogus Pars 118. *Coccinellidae I.* PP 224 *Berlin.*

Korschefsky, R., 1932. Coccinellidae II. In Junk-Schenkling *Coleopterorum Catalogus, Pars* 120 PP. 433, Berlin, Junk.

Korschefsky, R., 1933. Entomological investigations on the spike Disease of Sandal (16) Coccinellidae (Coleoptera). *Indian Forest Records*, 19(6), 1-17.

Kugelann, J.G., 1794. Verzeichniss der in eninigen Gegenden Preussens bis jetzt entdeckten Kaferarten nebst kurzen Nachrichten von denselben. *Neues Maqazin Liebh. Ent. Heft.*, 5 : 513-582.

Laing, D.R. and Caltgirone, L.E., 1969. Biology of *Habrobracon lineatellae. Can. Ent.*, 101 : 135-142.

Leach, W.E., 1815. (Articles on entomology). In Brewster, Edinburgh. *Encyclopaedie* 9 : 57-172.

Leconte, J.L., 1852. Remarks upon the Coccinellidae of the United States. *Proc. Acad. Natur. Sci. Philadelphia*, 6 : 129-145.

Leong, G.K.L. and Oatman, E.R., 1968. The biology of *Campoplex haywardi* (Hymenoptera : Ichneumonidae), a primary parasite of the potato tuber worm. *An. ent. Soc. Am.*, 61:26-36.

Lewis., G., 1873. Notes on Japanese Coccinellidae. *Ent. Mon. Mag.*, 10(2) : 54-56.

Lewis, G., 1879. On certain new species of Coleoptera from Japan *Ann. Mag. nat. Hist. Lond.*, (5)4 : 459-467.

Lewois, G., 1896. On the Coccinellidae of Japan. *Ann. Mag. nat. Hist. Lond.*, (6)17 : 22-41.

Lim Sook Ming, 1971. The biology and ecology of some aphidophagous Coccinellidae. Ph. D. Thesis. University of Malaya, Kaula Lumpur.

Linadauer, M., 1967. Recent advances in bee communication and orientation. *An. Rev. Ent.*, 12:439-470.

Linnaeus, C., 1758. Systema naturae per regna tria natureae, Secundum classes. ordines, genera, species, cum Characteribus, differentiis, Synonymis locis. Ed. 10, (1), PP 823. Homiae.

Lyons, L.A., Sullivan, C.R., Wallace, D.R. and Griffiths, K.J., 1972. Problems in the management of forest pest populations. *Proc. Tall. Timbers Conf.*, 3:129-140.

Mader, L., 1926-37. Evidenz der palaarktischen Coccinelliden and ihrer Abberrationen. Wort and Bild I. Teil., 412 + 12.

Mader, L., 1941. Coccinellidae, I Teil. PP. 136. *Eplor. Parc. nat. Albert. Miss. de. Witte fasc.* 34.

Mader, L., 1954. Coccinellidae, III Teil. pp. 206. *Explor. Parc. nat. Albert. Miss. de. Witte fasc.* 80.

Mader, L., 1955. Evident der palaarktischen Coccinelliden and ihrer Abberrationen. Wort and Bild I. Teil. *Entom. Arb. Mus. Frey*, 6 : 764-1035.

Manjunath, T.M., 1992. Biological Control from Lab to land : Roll of Commercial insectaries. *Proc. Nat. Workshop Biotech. Appro. Bio. Control Phytophaqus insects* pp. 32-33.

Markl., N. and Linadauer, M., 1965. Physiology of insect behaviour. PP 3-122 M.Rockstein, ed., The physiology of insecta, 2, Academic Press, New York.

Marriner, T.F., 1939. Movements of Coccinellidae. *Ent. Rec.*, 51 : 104-106.

Mayne, W.W., 1953. *Cryptolaemus montrouzieri* Mulsant in South India. *Nature*, 172 : 185.

McKenzie, H.L., 1932. The biology and feeding habits of *Hyperaspis lateralis* Mulsant (Coleoptera : Coccinellidae). *Univ. Calif. Publ. Ent.*, 6 : 9-20.

Miwa, Y. and Yoshida, T., 1935. Catalogue of Japanese Insects, 9. Coccinellidae. *Ent. World, Tokyo,* 3 (2) : 32-53.

Miyatake, M., 1961a. The East-Asian Coccinellid-beetles preserved in the California Academy of Science, tribe Hyperaspini. *Mem. Ehime Univ.,* (6)6 : 67-86.

Miyatake, M., 1961b. The East-Asian Coccinellid beetles peserved in the California Acaemy of Science, tribe Serangiini. *Mem. Ehime. Univ.,* (6)6 : 135-146.

Miyatake, M., 1961c. The East-Asian Coccinellid-beetle preserved in the California Academy of Science, tribe Hyperaspini. *Mem. Ehime. Univ.,* (6) 6 : 147-155.

Miyatake, M., 1967. Notes on some Coccinellidae from Nepal and Darjeeling District of India (Coleoptera). *Trans. Shikoku Ent. Soc.,* 9(3) : 69-78.

Miyatake, M., 1969. The genus *Plotina* and related genera (Coleoptera : Coccinellidae). *Pacific Insects,* 11(2) : 197-216.

Miyatake, M., 1970. The East-Asian Conccinellid-beetle preserved in the California Academy of Sciences, tribe Chilocorini. *Mem. Coll. Agr. Ehime. Univ.,* 14 (2) : 303-340.

Miyatake, M., 1985. Coccinellidae collected by Hokkaido University Expedition to Nepal, Himalaya, 1968 (Coleoptera). *Insecta matsumarana,* 30 : 1-33.

Motschulsky, V., 1866. Catalogue des insects recus du Japon. *Bull. Soc. Imp. Moscour.,* 39 : 164-200.

Moter, G., 1959. Untersuchungen zur Biologie von, *Stethorus ounctillum* Weise, Doktor-Dissertation, Mathem. *Nautuwiss. Fak. Univ. Koln.*

Moursi, A.A., Kamal, M., 1946. Notes on the biology and feeding habits of the introduced beneficial insect *Leis conformis* Boised. (Coccinell). *Bull. Soc. Fauad I. Ent.,* 30 : 63-74.

Muggleton, J., 1979. Non-random mating in wild populations of polymorphic *Adalia bipunctata. Heredity,* 42 : 57-65.

Muir, F., 1918. Notes on the ontogeny and morphology of the male genital tube in Coleoptera. *Trans. R. ent. Soc. Lond.* 1918 : 223-229 (W.L. 53980).

Mulsant, M.E., 1846. Securipalpus. PP. 380. Histoire Naturelle des Coleopteres de France, 4.

Mulsant, M.E., 1850. Species des Coleopteres trimeres Securipalpus. pp. 1104.

Mulsant, M.E., 1866. Monographic des Coccinellides Ire partie *Coccinellies* PP. 1-294. Paris.

Muma, M.H., Selhime, A.G. and Denmark, H.A., 1961. An annotated list of predators and parasites associated with insects and mites on Florida citrus. *Florida Aqri. Exp. Sta. Bull.*, 634 : 3-39.

Nagraja, H. And Hussainy, S.U., 1968. A study of six species of *Chilocorus* (Coleoptera : Coccineliidae) predaceous of San Jose and other scale insects. *Orient. Insects*, 1 : 239-241.

Nakahara, L., Magamine, W., Matayoshi, S. and Kumashiro, B., 1987. Biological control programme on the leucaena psyllid, *Heteropsylla cubana* Crawford. (Homoptera : Psyllidae) in Hawaii. *Leucaena Res.Rpt.* 7 : 39-44.

Nakane, T., 1959. Entomological results from the scientific survey of the Tokara Islands. (Coleoptera : Coccinellidae). *Sci. Rep. Kyoto. Pref. Univ.*, 3 : 45-52.

Nielsen, A., 1957. On the evolution of the genitalia in male insects. *Ent. Medt.* 28 (1) : 27-57.

Obata, S., 1986. Mechanisms of Prey findings in the aphidophagous lady bird beetle, *Harmonia axyridis* (Coleoptera : Coccinellidae). *Entomophaqa*, 31 : 303-331.

Obata, S., 1987. Mating behaviour and sperm transfer in the lady bird beetle, *Harmonia axyridis* Pallas (Coleoptera : Coccinellidae). *Appl. Ent. Zool.*, 22(4) : 434-442.

Obata, Shohko and Hidaka, Toshitaka, 1987. Ejection and ingestion of the spermatophore by the female lady bird beetle, *harmonia axyridis* Pallas (Coleoptera : Coccinellidae). *Can. Ent.*, 603-604.

O' Donald, P. and Muggletoin, J., 1979. Melanic polymorphism in lady birds maintained by sexual selection. *Heredity*, 43 : 57-65.

Omkar and Bind, R.B., 1993. Records of Aphid natural enemies

Complex of Uttar Pradesh. II. The Coccinelids. *J. Adv. Zool.*, 14(2) : 96-99.

Pajni, H.R. and Sigh, J., 1982. A report on the family Coccinellidae of Chandigarh and its surrounding areas (Coleoptera). *Res. Bull. Punjab Univ. Sci.*, 33 : 79-86.

Pajni, H.R. and Verma, S., 1985. Studies on the structure of the male genitalia in some Indian Coccinellidae (Coleoptera). *Res. Bull. Punjab Univ. Sci.*, 36 : 195-201.

Palmer, M.A., 1914. Some notes on life history of lady beetles. *Ann. ent. Soc. Am.*, 7 : 213-238.

Paroda, R.B., 1993. Crop science research for sustainable agriculture in India. *proc. Nat. Acad. Sci. India* (B), 63 : 97-114.

Patro, B.and Behera, M.K., 1992. Development and feeding potential of *Scymnus* sp. on the Bean Aphid, *Aphis craccivora*. *J. Biol. Control*, 6(2), 64-66.

Putman, W.L., 1955. The immature stages of *Stethorus punctillum* Weise (Coleoptera : Coccinellidae). *Can. Ent.*, 86 : 506-508.

Putman, W.L., 1957. Laboratory studies on the food of some coccinellids (Coleoptera) found in Ontario peach orchards. *Can. Ent.*, 89 : 572-579.

Pattarudraiah, M. and Channabasavanna, G.P., 1952. Discovery of *Cryptolaemus montrouzieri* Mulsant (Coccinellidae : Coleoptera) in Bangalore, South India. *Nature*, 169 : 377-378.

Pattarudraiah, M. and Channabasavanna, G.P., 1953. Beneficial Coccinellids of Mysore - I. *Indian J. Ent.*, 15 : 87-96. (W.L. 22997).

Puttarudraih, M. and Channabasavanna, G.P., 1955. Beneficial coccinellids of Mysore -II. *Indian J. Ent.*, 17 : 1-5.

Puttarudraiah, M. and Channabasavanna, G.P., 1956. Some beneficial Coccinellids of Mysore-III. *J. Bombay. Nat. Hist. Soc.*, 54 : 156-159.

Rafiq Ahmad, 1973. A new tribe of the family Coccinellidae (Coleoptera). *Bull. ent. Res.*, 62 : 449-452.

Rao, T.V. and David, A.L., 1958. The biological Control of Coccid.

pest in South India by the use of the beetle, *Cryptolaemus montrouzieri* Muls. *Indian. J. agric. Sci.*, 28(4) : 545-552.

Rao, V.P., 1971. A review of the biological control of insects and other pests in South East Asia and the Pacific region. pp 149. Commonwealth Agricultural Bureaux, Farnham Royal, Slough SL2 3BN, England.

Rao, C.S, Rao S.V. and Rao, B.H., 1981. Biology of the Exotic Predator *Chilocorus cacti,* (Coccinellidae : Coleoptera) on sugarcane scale Insect *Melanaspis glomerata* Green. *Cooperative sugar.* 12, (6) : 303-304.

Raulston, J.R., Graham, H.M., Lingren, P.D. and Snow, J.W. 1976. Mating interaction of native and laboratory reared tobacco bud worms released in the field. *Enviorn. Entomol.*, 5:195-198.

Rawat, U.S., Sangal, S.K. and Pawar, A.D., 1992. Biology of *Chilocorus bijuqus* Mulsant (Coleoptera : Coccinellidae), Predator of San Jose scale, *Quadraspidiotus perniciosus* (Comstock). *J. Biol. Control,* 6(2), 97-100.

Redtenbacher, L. 1843. Tentamen dispositionis generum et specierum Coleopterorum Pseudotrimerorum Archiducatus Astriae, reprint 1844. *Zeiteschr. Entomol.*, 5 : 113-132.

Rees, B.E., 1947. Taxonomy of the larvae of some North American Noviini. *Pan-Pacific Entomol.*, 23 (1) : 113-119.

Roeder, K.D., 1935. An experimental analysis of the sexual behaviour of the praying mantis (*Mantis religiosa* L.). *Biol. Bull.*, 69:203-220.

Sasaji, H., 1967. On the Coccinellidae attacking the aphids in Japan and Ryukyus. *Mushi*, 40 : 147-175.

Sasaji, H., 1968a. Phylogeny of the family Coccinellidae (Coleoptera). *Etizenia* 35 : 1-37.

Sasaji, H., 1968b. Description of the Coccinellid larvae of Japan and the Ryukus (Coleoptera). *Mem. Fac. Edn. Fukui Univ.* 2, (18) : 93-136.

Sasaji, H., 1970. Biogeographical consideration of the Coccinellidae of the Tsushimas, Japan (Coleoptera). *Bull. Nat. Sci. Mus. Tokoyo.*, 13 : 206-217.

Sasaji, H., 1971. Fauna Japonica Coccinellidae (Coleoptera). pp. 340. Academic Press, Japan.

Sathe, T.V., 1985. Studies on mating, oviposition and emergence of *Cotesia diurnii* Rao and Nikam (Hymenoptera : Braconidae) an internal larval parasitoid of *Exelstis atomosa* Walsingham. *Geobios*, 41, 100-101.

Sathe, T.V., 1986. Biology of *Cotesia diurnii* R and N (Hym. : Braconidae) a larval parasitoid of *Exelastis atomosa* Walsingham. *Oikoassay*, 3 : 31-33.

Sathe, T.V. 1981. Crop protection from insects pp. 70. Ed. I, Bhiungade Press, Kolhapur.

Sathe, T.V., 1992. Fauna of aphids on plants of economic importance found in Western Maharashtra. *J. Curr. Biosci.*, 9 (1) : 27-31.

Sathe, T.V. and Nikam, P.K., 1983. Adult longevity of *Cotesia flavipes* (Cameron) (Hymenoptera : braonidae), with different food. *Sci. & Cult.*, 49, 405-406.

Sathe, T.V. and Nikam, P.K., 1984. Mating, oviposition and emergence of *Cotesia orientalis* C. and N., (Hymenoptera : Braconidae), an internal larval parasitoid of *Exelastis atomosa* Walsingham. *Camp. physiol. Ecol.* 9, 231-232.

Savoiskaya, G.I., 1955. Saxaulical ladybird - *Brumus jacobsoni* Bar. (Col., Coccinellidae). *Uchem. Zap. biol. poclv. Fak. Kirqiz. Univ.*, 5 : 319-326.

Savoiskaya, G.I., 1957. The experience of composition of key of Coccinellids larvae. *Dokl. VII nauch. Knof. Tomsk. Univ.*, 3 : 62-63.

Savojskaja, G.I. (Savoiskaya, G.I.), 1960. On morphology and taxonomy of lady birds larvae (Coleoptera, Coccinellidae from South-east Kazakistan. *Ent. Obozr.*, 29 : 122-133.

Sem'yanov, V.P., 1970. Biological properties of *Adalia bipunctata* L. (Coleoptera : Coccinellidae) in condition of Leningrad region. *Zasch. Rast. Vredit. Bolez.*, 127 : 105-112.

Shull, A.F., 1948. Natural hybrids of sub species of *Hippodamia quainquesignata*. *Evolution*, 2, 10-18.

Sicard, A., 1907. Coleopteres Coccinellides du Japon. *Bull. Mus. hist. nat. Paris.*, 1907 : 211.

Sicard, A., 1909. Revision des Coccinellides de la faune. *Ann. Soc. ent. Fr.*, 78 : 68-165.

Simmonds, F.J., 1963. Genetics and biological control problems. *Entomophaga*, 17:251-264.

Simmonds, F.J., 1970. Biological Control. Part 4/8. Sampond Press Ltd. Bangkok.

Singh, A. And Sadhu, A.N., 1986. Agricultural problem in India. pp. 386. Jammu University, Jammu.

Singh, R., Malhotra, R.K., 1979. Some studies on the biology of *Coccinella undecimpunctata memetriesi* Muls., a predator of Mustard aphid. *Curr. Sci.*, 48 (20) : 904-405.

Strouhal, H., 1926. Die Larven der Palaerktischen Coccinellini and Psylloborini (Coleopt.). *Arch. naturqesch.*, 92(A), 3 : 1-63.

Subramaniam, T.V., 1923. Some coccinellids of South India. PP. 108-118 *Rep. Proc. Fifth Ent. Meeting, Pusa.*

Subramaniam, V.K., 1955. Control of the fluted scale in Peninsular India *Indian J. Ent.*, 16(1954) : 39-416; 17(1) : 103-120.

Sweetman, H.L., 1958. The principles of Biological control. *Bubuque; Wm. C. Brown.*, 8.

Tawfik, M.F.S., Salah Abul Nasar & Botha Una M. Sadd, 1973. The biology of *Scymnus interrupts* Goeze (Coleotptera : Coccinellidae) *Bull. Soc. Entomol Egypt.*, 57 : 4-26.

Timberlake, P.H., 1919. Revision of the Parasitic Chalcidoid flies of the genera *Homalotylus* Mayr. and *Isodromus* Howard, with description of two closely related genera. *Proc. U.S. Nat. Mus.*, 56 : 133-194.

Timberlake, P.H., 1943. The Coccinellidae or lady beetles of the Koebele Collection Part II, Hawaii. *Plant Rec.*, 47 : 1-67.

Usman, S. and Puttarudriah, M., 1955. *A list of the insects of Mysore including the mites.* PP. 194. *Ent. Ser. Bull.*, No. 16, *Dept. Aqric.*

Watson, J.R., Thompson, W.L., 1933. Food habits of *Leis conformis* Boisd, (Chinese lady beetle). *Fla. Ent.*, 17 : 27-29.

Watson, W.Y., 1956. A study of the phylogeny of the genera of the tribe Coccinellini (Coleoptera). *Centr. Roy. Ont. Mus. Toronto* (*Zool.*), 42 : 1-52.

Weise, J., 1885. Best immungstabellen der europasischen Coleopteren. PP. 83. Coccinelliden, 2. Aufl.

Weise, J., 1900. Coccinelliden aus Ceylon gesammelt von Dr. Horn. *Dtsch. ent. Z.*, 1900 : 417-445.

Williams, J.I., 1945. The anatomy of the internal genitalia of some Coleoptera. *Proc. ent. Soc. Washington*, 47 : 73-87.

Wilson, J.W., 1926. The genitalia of some of the Coccinellidae. *J. Elisha Mitchell Sci. Soc.*, 1 : 63-74.

Wingo, C.W., 1952. The Coccinellidae (Coleoptera) of the upper Mississippi basin. Iowa State. *Coll. J. Sci.* 27 : 15-53.

Wu, C.F., 1937. Coccinellidae. *Catalogues Insectorum. Sinensium* 3(2) : 553-587.

Wylie, H.G., 1958. Observations on *Aphidecta obliterata* (L.) (Coleoptera : Coccinellidae) a predator of conifer infesting Aphidoidea. *Can. Ent.*, 90 : 518-522.

Yadava, C., Shaw, F. R., 1968. The preference of certain Coccinellids for pea aphids, leafhoppers and alfalfa larvae. *J. econ. Ent.*, 61 : 1104-1105.

Zacharuk, R.Y., 1958. Structures and functions of the reproductive systems of the prairie grainwire worm, *Ctenicera aeripennis destructor* (Brown) (Coleoptera : Elater idae). *Can. J. Zool.*, 36 : 721-751. (W.L. 13191).

Zaslavskij, V.A., 1965. Fam. Coccinellidae. [key to the Insect of European region of U.S.S.R.]. 2 : 319-326. (*In Russian*).

Subject Index

Species Index